Green Networking

Green Networking

Edited by
Francine Krief

First published 2012 in Great Britain and the United States by ISTE Ltd and John Wiley & Sons, Inc.

ISTE Ltd
27-37 St George's Road
London SW19 4EU
UK

www.iste.co.uk

John Wiley & Sons, Inc.
111 River Street
Hoboken, NJ 07030
USA

www.wiley.com

Library of Congress Cataloging-in-Publication Data

Green networking / edited by Francine Krief.
    pages cm
  Includes bibliographical references and index.
  ISBN 978-1-84821-378-4
    1. Telecommunication--Energy conservation. 2. Telecommunication--Environmental aspects. 3. Computer networks--Environonmental aspects. 4. Sustainable engineering. 5. Computer networks--Energy conservation. 6. Green technology. I. Krief, Francine.
  TK5102.5.G734 2012
  384.028'6--dc23

2012028169

British Library Cataloguing-in-Publication Data
A CIP record for this book is available from the British Library
ISBN: 978-1-84821-378-4

Printed and bound in Great Britain by CPI Group (UK) Ltd., Croydon, Surrey CR0 4YY

Table of Contents

**Chapter 9. Industrial Application of Green
Networking: Smarter Cities** 233
Vincent GAY, Paolo MEDAGLIANI, Florian BROEKAERT,
Jérémie LEGUAY and Mario LOPEZ RAMOS

Introduction

Information and Communication Technology (ICT) is increasingly omnipresent in most human activities, and is viewed primarily in terms of its contribution to economic productivity and to our wellbeing. However, with the necessity to reduce our overall CO_2 emissions to protect our environment, and the constant increase in the cost of energy, the carbon footprint of these technologies has become a cause for concern. Today, the energy impact of ICT is evaluated at 2% of greenhouse gas emissions (equivalent to that of all aviation the world over); in heavily industrialized countries, this figure can reach up to 10%. In addition, the energy consumed by ICT is increasing by about 15-20% each year. The development of new types of applications such as HDTV, new usages such as ubiquitous networks and the explosion in the volume of data traffic on 3G/3G+, and soon to be 4G/LTE networks, would suggest that this consumption is not likely to slow down any time soon.

However, at present, energy consumption is far from being optimized. Communication networks are usually over-dimensioned and designed with redundant capacity. Numerous networking devices consume a considerable amount of energy, even when they are not being heavily used, or not used at all; for instance, this is the case with the Base Transceiver Stations in cellular networks.

The concept of green networking has recently emerged. This encapsulates all approaches and measures employed to reduce the volume of greenhouse gas emissions due to the process of communication.

The objective of this book is to offer an overview of the mechanisms and procedures used for implementing energy-efficient networks and limiting their carbon footprint. Some of these devices are already operating – particularly in mobile networks; others are near the point of becoming operational; finally, some proposals and promising directions for future research are presented.

This book, which is one of the first ever to present a "state of the art" on the advances and research projects in the domain of green networking is made up of nine chapters, and is structured in three parts.

Chapter 1 introduces the problem of reducing the electricity consumption of ICT and particularly for telecommunications infrastructures, given that their CO_2 emissions are increasing greatly.

The next three chapters discuss the energy efficiency of communication networks, with each chapter focusing on a particular technology. Together, they make up *Part 1* of this book, entitled *"A Step Towards Energy-efficient Networks"*.

Chapter 2 looks at energy gains in operational wired networks, meaning networks for which the design phase has been completed and the infrastructure is already in place. These communication networks are usually over-dimensioned and built with redundant capacity. For that reason, the opportunities they represent for making energy savings are considerable. Several strategies for optimizing energy consumption are presented in this chapter – in particular, those that lend themselves to energy-efficient routing – which are then evaluated on the basis of "real-world" scenarios.

While the model of the consumption of wired communication links remains rather imprecise, the same is absolutely not true for radio links, which constitute an undeniable hotbed of optimization opportunities, because the processing of a radiofrequency signal is a very energy-hungry operation. In addition, of the various actors involved in telecommunications networks, mobile network operators are the main consumers of energy today.

Chapter 3 discusses the environmental impact of mobile networks and the necessity to reduce their energy consumption. Its primary focus is on the techniques employed to reduce power in the radio access network, given that this is the part which consumes most energy. Two other axes for optimization to reduce greenhouse gas emissions are then presented: the first relates to the architecture and engineering aspects of mobile networks, and the second to the components and structures of these networks.

Chapter 4 introduces the energy consumption of the data centers which form the memory of the Internet, and increasingly its computing power and applications. The focus then moves onto low-cost access networks with low energy consumption. Two solutions would seem to suggest themselves here: femtocells and mesh networks. Finally, the chapter highlights virtualization techniques, which facilitate more appropriate multiplexing and the shutdown of needless hardware resources.

The next three chapters deal with the contribution of new technologies to improve the energy efficiency of networks. They constitute *Part 2* of this book, entitled *"A Step Towards Smart Green Networks and Sustainable Terminals"*. The issue is the anticipated gains, which will become greater still as these new technologies bring more users into the fold.

Chapter 5 discusses cognitive radio networks, which are emerging as a new concept in wireless communications, capable of dealing with the lack of radio resources. Thanks to its agility and capacity to intelligently adapt the parameters of communication, cognitive radio can be exploited to render wireless devices more energy efficient. Several possibilities for exploration are proposed in this chapter.

Chapter 6 applies the concept of autonomic networking to green networks. Green networks would then become capable of self-organization and self-adaptation in order to maintain efficient and environmentally-friendly operation, even in the presence of changeable conditions. Firstly, the four main self-functions are introduced – these are self-configuring, self-optimizing, self-protecting and self-healing. Their respective contributions in the context of green networks are later described and illustrated, taking the particular examples of wireless sensor networks, which by nature are energy-constrained, and smart grids, which contribute to decreasing greenhouse gas emissions.

Chapter 7 studies the environmental impact of communication terminals throughout their lifecycle. This impact, while it is less than that of other industries such as transport, is far from negligible when we consider the particularly high rate of replacement of mobile telephones and smartphones, which is linked to the problem of electronic waste. In addition, in the prediction of a digital society which is more respectful of the environment, it is essential to reduce the environmental impact of electronic products, which will occupy a phenomenally important place in that society in the future. This impact decreases greatly with a longer lifespan of the product. This chapter proposes an interesting avenue for improvement: to design reconfigurable hardware systems in order to increase their functional lifespans. Today, this solution is feasible, thanks in particular to the advent of reconfigurable hardware circuits, and offers better recycling.

The *final part* of this book, entitled *"Research Projects on Green Networking Conducted by Industrial Actors"*, is made up of two chapters. The first has been written by a telecoms operator, and the second by a large group specialized in products and systems of ITC.

In the context of the struggle against global warming, mobile operators are looking for ways to reduce the energy consumption of their equipment. However, this reduction must not impact the Quality of Service offered to the customers.

Chapter 8 describes research projects relating to mechanisms for putting Base Transceiver Stations in sleep mode, and their impacts on the overall consumption of mobile networks. This is one of the flagship solutions which drastically reduces energy consumption. Sleep mode is applied to various scenarios for deployment of networks: a conventional cellular network and a heterogeneous network made up of small cells. Sleep mode can offer significant gains in certain use cases, whilst still preserving the Quality of Service perceived by the user.

The smart city can be viewed as an industrial application of green networking.

Chapter 9 presents the concept of a smart city, which involves developing new generations of urban infrastructures heavily influenced by green networking. Indeed, numerous devices will be interconnected – particularly sensors, actuators, video cameras, Base Transceiver Stations, data servers, PCs for command centers, users' smartphones, etc. The usage of these resources must be optimized by using energy-efficient techniques. This chapter gives an overview of different aspects of green networking which are able to reduce the number or the consumption of the devices produced and the networks put in place: low-consumption communication protocols, assistance with the deployment of a wireless sensor network, low-consumption processor treatments and finally, integration and use of sensors to help in deciding on energy-saving policies. The chapter closes with an example of the use of these technologies to respond to the need for energy management.

Problems linked to green networking are at the heart of research in the field of networks today, and of the societal stakes we find ourselves facing. A great deal of work has still to be done to reduce the consumption of the protocols, communication architectures and networking equipment without damaging the Quality of Service and security.

Taking account of the carbon footprint has to be a constant concern of R&D engineers in telecommunications. However, while ICTs do consume energy, they can also contribute to a reduction in our global CO_2 emissions, e.g. by limiting our movements and optimizing the energy consumption of our dwellings and cities.

Chapter 1

Environmental Impact of Networking Infrastructures

1.1. Introduction

Over the past decade, the issues relating to the cost of information and communication technology (ICT) have increased considerably. These issues stem from a number of disciplines: ecology, economics, politics and societal matters. "Green IT" encapsulates efforts taken to reduce the energy footprint related to ICT, or at the very least to slow its growth.

The environmental cost of ICT is a hot topic, because it is highly controversial. The most alarming estimations associate an impact of around 10% of electricity worldwide for ICT, and 2% of global energy, which is constantly increasing. These measurements are backed up, and projected to increase by around 10% a year over the next ten years [EPA 07]. In France, energy consumption by ICT is estimated to be between 55 and 60 TWh per year, which equates to 13.5% of electricity consumption by the end

Chapter written by Laurent LEFÈVRE and Jean-Marc PIERSON.

applications [TIC 08]. To express this as a financial cost, we must remember that the price of 1 kWh in France is €0.10, which is equivalent to around €900,000 for 1 megawatt/year. Yet these figures only take account of the costs of direct usage and the electricity impact, overlooking the aspects relating to the production, transport and recycling of the products.

As the GeSI (*Global e-Sustainability Initiative*) has shown (see Figure 1.1), the constantly-increasing trend is such that if it continues, in 2020 ICT will produce 1.43 Gtons of CO_2, which represents 2.7% of the total carbon footprint, distributed between usage (1.08 Gtons) and the rest (production, transport, recycling – 0.35 Gtons) [GES].

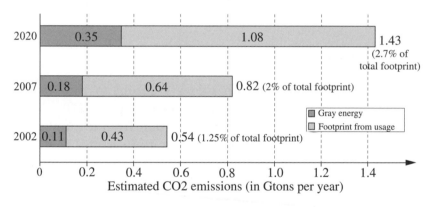

Figure 1.1. *Evolution in CO_2 emissions from ICT: 2002-2020 [GESI]*

It is a necessity to develop an ICT sector which is sustainable for the planet. Indeed, it is inexplicable that a sector as omnipresent as this in the daily lives of citizens should not be concerned with reducing CO_2 emissions, controlling energy, and making a firm commitment to collective responsibility. The millions of subscribers to communication and entertainment services using ICT are responsible, and must become active participants in the changes made in (and for) the future. Every connection to an

Internet network, every usage of a social network, every request sent to a search engine and every video watched entails a cost – a cost which it is helpful to understand and lay bare, in order to be able to reduce it later on.

Behind the powerful tools and services which today's society cannot go without, there is a massive infrastructure of processing and communication: the platforms of multinational players in Internet and social networks, and of the banking systems, contain millions of machines distributed between data centers, sometimes organized into a "cloud"; the planet's communication networks, comprising both wired and wireless technologies (fiber optic, copper, satellite, WiFi, GSM) provide a constantly-increasing throughput of communication to channel texts, images, videos, etc. – ultimately, *bits* of information. If the predictions of a 60% annual growth in Internet traffic, the majority of which comes from the entertainment sector (online games, higher-resolution audio/video jukeboxes, etc.), are correct, by 2025 there will be a transatlantic traffic of more than 400 Tbit/s.

Network operators are among the most electricity-hungry players in the industry. In 2011, Telecom Italia [TEL] estimated its consumption to represent 1% of Italy's total electricity consumption (in comparison to 0.7% in 2008). Similarly, British Telecom estimates its share of the UK's electricity consumption to be 0.7% (2.3 TWh) – a proportion similar to that declared by NTT in Japan. These figures take account of all sites where electricity is expended – from the headquarters to the network infrastructure and the data centers associated with it. For instance, for Telecom Italia's infrastructure, 65% of the electricity is consumed by the networks (both landline and mobile) and 10% by the data centers. However, the figures do not include consumption by the equipment in the premises of the end users. In France, a study conducted by IDATE [IDA] states the overall energy consumption of the telecoms sector at 8.5 TWh in 2012 (by contrast to 6.7 TWh in 2008). This consumption is distributed between the infrastructure (wired and mobile networks: 46% of which data centers accounted for 6% in 2008), the ADSL boxes in users' homes (24%) and both hardwired and wireless telephones (18 %), also in the users' premises. Note that the total consumption of domestic routers in France is estimated at 3.3 TWh in 2012 (40 million boxes).

CO_2 emissions due to communications networks are also increasing greatly, as shown by another analysis carried out by the GeSI (Figure 1.2). The part played by broadband networks has burgeoned (from 3% to 14%), the mobile network increases less in terms of percentage (from 43% to 51%), but doubles in terms of its absolute value, whereas the contribution of peripheral infrastructure increases from 12 to 20%. Only narrow band networks are expected to see their emissions decrease in the future, disappearing little-by-little and improving their efficiency to the benefit of other types of network. According to IDATE, every European mobile phone user is responsible for the emission of 17 kg of CO_2 every year, while landline and Internet users are responsible for 44 kg of CO_2.

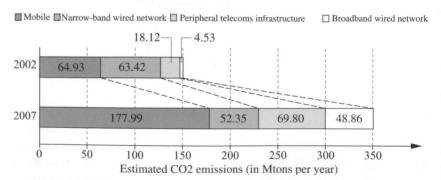

Figure 1.2. *CO_2 emissions due to networks*

1.2. Some definitions and metrics

When we speak of the energy consumed (by IT equipment), this is the product of the usage time (in seconds) by the instantaneous power (in Watts). Also, energy is expressed in W.s, or in Joules (1 Joule = 1 W.s). From this definition, we can already see that reducing the energy consumed involves either decreasing the time (going faster) or decreasing the power (improving the efficiency of the components and their use).

We consider that there are two types of energy consumed by ICT equipment (servers, storage, networks, etc.):

– static energy: energy consumed by the "idle" power supply to the equipment (for instance, a router which is not carrying any data at the time, a server which is not performing any service, etc.);

– dynamic energy: that proportion of the energy consumed by the machine when it is in use.

Schematically, ICT infrastructures are made up of two levels: a hardware level and a software level. Thus, when we are interested in their environmental impact, it is helpful to look, at both these levels, at what can be measured, calculated, improved, sometimes considering the two levels independently, and sometimes combining them.

In order to be able to compare software and hardware infrastructures, we have to define a "meter stick" of sorts – a universal standard by which to measure: the energy consumed is one such meter stick, but it is not enough. Obviously, we have to take account of the work performed by the infrastructure in question, whether it be a computation or communication infrastructure.

Numerous metrics have been put forward and are still being put forward today, which demonstrates that the research discipline which deals with green IT is still in its infancy.

Here, we shall cite some of the metrics that have been proposed, but with no pretense at exhaustivity:

– FLOPS per Watt represents the peak power in terms of operations per second (1 Flops = 1 FLOating Point operation per Second), expressed per unit power. This metric measures the peak capacities of a machine and is particularly useful for supercomputers and the world of high-performance computation, e.g. in the Green500 classification [GRE a];

– Joules/bit represents the energy needed to process one bit of information. This metric can serve for computation, storage and communication. Thus, in networking equipment, for each component we can distinguish the energy needed to process and transfer one bit of information;

– PUE (Power Usage Effectiveness) represents the energy efficiency of computation and storage infrastructures such as Clusters and Clouds (data centers in general). This is the ratio between the power injected into the data center and the power used by the machines (thus eliminating energy losses, cooling etc.). Contemporary data centers have a PUE between 1.1 for the best, and 2 for the least efficient. This metric includes the computation, storage and communication aspects for the part measuring the power of the machines;

– CUE (Carbon Usage Effectiveness) considers the impact in terms of CO_2 emissions. By contrast with the previous metric, this takes account of the type of energy by which the data center is supplied. It is measured in kilograms of CO_2 per kilowatt hour. It is interesting that this is one of the "scarce approaches" which take account of the environmental impact beyond the quantity of electricity consumed. Similarly, metrics such as WUE (Water Usage Effectiveness) and ERE (Energy Reuse Effectiveness) amongst others enable us to compare the infrastructures based on their environmental aspects. Note that, to date, no metric integrating the production and recycling of the components has become available. These metrics (PUE, CUE, ERE, etc.) were presented by the Green Grid consortium [GRE b].

In what follows, our primary focus is on the infrastructures of interconnecting networks. Hence, the metrics which we shall use are mostly related to the energy *per se*, and to the energy expressed as a function of the number of bits processed.

1.3. State of the sites of consumption of the networks: the case of wired networks

Evaluating the energy consumption of an interconnecting network is no easy task: we have to coordinate a set of measurements of the instantaneous power consumed by each component of the network and its activity time. It is a challenge to measure this power: the quality of the measurement, its precision and its frequency depend on the measuring devices used. Indeed, the power measurement recorded by an external wattmeter measuring the whole of a device, and the measurement recorded component-by-component, will give neither the same degree of precision nor the same quality. Both types of evaluation are to be found in the existing body of literature.

A still more difficult task is to measure the consumption caused by a given particular communication, because we have to take account of the sharing of resources in equipment which is able to process several data flows at the same time.

Studies and projections concerning the energy consumption of networks are few and far between. In 2008, Asami *et al.* [ASA 08], studying the case of Japan, projected that, even if low-consumption electronic components were used, the energy consumption of IP routers would, in the 2030s, exceed Japan's total energy-producing capacity (such as it was in 2005).

In 2010, Zhang *et al.* [ZHA 10] looked at optical networks. Taking the figures from 2009 as a basis, the authors estimated that the energy consumption of optical networks will increase by 120% ($\times 2.2$) between now and 2017.

Bolla *et al.* [BOL 10] estimated the overall consumption of wired networks in Italy between 2015 and 2020. Their results, summarized in Table 1.1, show that the part of the network in users' homes represents in total 79% of overall energy consumption (1,947 GWh/year), for 17.5 million access points. This fundamental study clearly shows where efforts in terms of research and awareness-raising are needed most desperately.

	Consumption per machine (in W)	Number of machines	% of overall consumption
Home	10	17,500 000	79%
Access network	1,280	27,344	15%
Metro/transport	6,000	1,750	< 5%
Core network	10,000	175	< 1%

Table 1.1. *Percentage of energy consumption for the different types of networks*

In 2008, Tucker *et al.* [TUC 08] measured the power consumed by IP routers depending on the technology used in these routers (Mb/s, Gb/s, Tb/s). From this, they deduced a mathematical law, linking the power consumed (P) in Watts to the capacity of the link (C) expressed in Mb/s: $P = C2/3$. A 1 Tb/s router consumes around 10,000 W, and a 1 Gb/s router consumes 100 W. Thus, an energy injection of 100 nJ per bit is needed in a 1 Gb/s router, whereas only 10 nJ per bit are needed in a 1 Tb/s router.

In the same study, the authors focus on optical routers, analyzing the energy needed for the various components of such a router. This study comes in the wake of a previous article [TUC 07] about *optical switches*. In both cases, we note that:

– the proportion of energy attributable to electrical supply and cooling is around 35%;

– the proportion due to the control level (mainly updating of the routing tables) is 10%;

– the part due to the data level (decoding of the IP header, IP transfer, inputs/outputs, buffers, etc.) represents 55% of energy consumption;

– these percentages do not change whether the technology used is entirely optical or electronic, and in time the difference between optical and electronic technology will decrease.

Given that the largest part of the consumption is due to the processing of the header, it is clearly advantageous to work on this point – e.g. by decreasing the number of hops, decreasing the work in the network core, privileging data flows passing through without staying for any length of time on intermediary nodes, etc.

The energy efficiency of equipment based on CMOS technology increases by a multiplicative factor of 1.65 every 18 months (Dennard's scaling law). In [KOO 11], the study over decades of energy efficiency shows that pieces of equipment such as servers' efficiency doubles every 1.57 years: the number of computations per Joule continued to double at this pace between 1949 and 2010. Expressed differently, a hundred-fold decrease is/will be observed in the cost in energy for a fixed workload every ten years. At the same time, processing capacities increase a hundredfold every 10 or 11 years. Ultimately, therefore, the individual electrical power of the machines has remained fairly constant over the past 60 years [AEB 11].

In [BOL 11], the authors show the increase in power of networking machines in view of their performance (see Figure 1.3). While their capacity (measured in Mbits processed per second) was multiplied by 2.5 every 18 months, their energy is multiplied by 1.65 over that same time period.

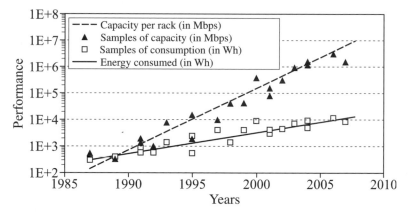

Figure 1.3. *Energy and capacity of routers over time*

The traffic predictions associated with networks are difficult to evaluate. Recently, a report from [CISCO] mentioned an explosion in traffic (by 2020), with this traffic being generated by new types of application (HDTV) and new uses (mobile and ubiquitous networks).

Currently, a great many pieces of networking equipment consume a considerable amount of energy even when they are hardly being used (if at all). Thus, the static part of their consumption is very significant in comparison to the dynamic part. The design of network components whose electrical consumption is proportional to their use (traffic, workload, volume of data exchanged, etc.) is one of the ambitious objectives of researchers currently working in this domain.

1.4. Academic and industrial initiatives

For a number of years, governmental, intergovernmental and industrial initiatives have been set up in order to pool "live" strengths on the subject. This section details some of the most noteworthy initiatives in terms of reducing the energy footprint of networks.

One of the most ambitious initiatives is spearheaded by the GreenTouch consortium [GRE d]. GreenTouch is aiming, by 2015, to make recommendations of advances to reduce the energy consumption of networks by a factor of 1,000. This reduction is accompanied by projections of the Quality of Service (QoS) and traffic support in the future. Structured into working groups (focusing on optical core networks, routing and switches, wireless mobile communications, access networks, and services), the consortium draws on the support of academics and industrial actors for approaches which combine hardware and software aspects. A set of focused projects, demonstrators and prototypes (both software and hardware) validate GreenTouch's proposals.

TREND [TRE] is a "network of excellence" on energy-efficient networking. Funded by the European Commission's FP7 research program, this cooperative of 12 research centers and industrial actors seeks to quantitatively measure the current and future demand for telecoms infrastructures and design sustainable networks on a reduced scale. Note that a great many of the figures relating to the consumption of networks cited in this chapter are taken from TREND.

The European project ECONET (*low Energy COnsumption NETworks*) [ECO] focuses on reducing the energy of wireline networks by favoring a set of dynamic technologies (sleep mode and performance adaptation). The objective is to reduce the electrical consumption of network equipment by 50% in the medium term and 80% in the longer term, while preserving the same level of end-to-end Quality of Service.

The COST Action IC0804 [ICO] is an open collaborative action financed by the European program COST as part of FP7. The action focuses on the energy efficiency of large-scale systems. Two specialized working groups on wired networks and wireless networks bring together European researchers working on these topics [PIE 10; PIE 11].

The European project PrimeEnergyIT (2009-2012) [PRI], financed by *Intelligent Energy in Europe*, is exploring the various technologies (servers, networks, storage, cooling), metrics and test banks, and certifications related to small and medium-scale data centers. In order to encourage thinking about green hardware from the stage of the very design of the data centers, PrimeEnergyIT provides a set of recommendations aimed at those who hold the European public spending purse strings on data centers and computation. PrimeEnergyIT also makes a variety of educational materials freely available.

The Canadian initiative GreenStar [GRE c] proposes to set up a medium-scale experimental network using only "green" energy (solar or wind power). This network links various academic centers the world over, and operates in "best effort" mode: if the energy production conditions (sunshine/wind) allow it, the network is available and transports information. If not, the machines of the network enter sleep mode. An adapted software environment is made available to its users.

1.5. Perspectives and reflections on the future

The reduction of electrical consumption of communication networks can be viewed from different angles: financial, philosophical or environmental. This reduction is part of a more general context, which is that of the effort to limit the use of resources (be they fossil, nuclear or green) so as to reduce the generation of greenhouse gases. Certain researchers believe that the approach chosen to improve energy efficiency is not aggressive enough [BIL]. In any case, the human race is going to have to face a significant increase in global temperature, and we must prepare for that. It seems inevitable that the human factor cannot be ignored in this globalizing approach, and that new Quality of Service paradigms must be put at users' disposal. "Green networking", and more generally green IT, must be taken as a necessary innovative factor for research organizations and enterprises [HER 12].

In a globalized green networking approach, it seems probable that "energy-hungry" technological solutions will give way to more energy-efficient solutions (e.g. DSL in comparison to optical networks). In addition, closer interaction between energy providers and massive consumers seems indispensable in order to balance supply and demand. Thus, many researchers see the proposals for "smart grids" as a future axis for green, energy-efficient networks.

1.6. Bibliography

[AEB 11] AEBISCHER B., "ICT and Energy", *ICT for a Global Sustainable Future Symposium*, http://www.cepe.ethz.ch/publications/Aebischer_Paradisio_ClubofRome_15-12-11_14-12-11.pdf, December 2011.

[ASA 08] ASAMI T., NAMIKI S., "Energy Consumption Targets for Network Systems", *ECOC 2008*, Brussels, Belgium, September 2008.

[BOL 10] BOLLA R., BRUSCHI R., CHRISTENSEN K., CUCCHIETTI F., DAVOLI F., SINGH S., "The potential impact of green technologies in next-generation wireline networks: is there room for energy saving optimization?", *IEEE Communication Magazine*, November 2010.

[BOL 11] BOLLA R., BRUSCHI R., DAVOLI F., CUCCHIETTI F., "Energy efficiency in the future internet: a survey of existing approaches and trends in energy-aware fixed network infrastructures", *IEEE Communications Surveys and Tutorials (COMST)*, 13 (2), May 2011.

[CIS] CISCO, Cisco visual networking index: Forecast and methodology, 2010-2015, 1 June 2011.

[GES] GLOBAL E-SUSTAINIBILITY INITIATIVE (GeSI), SMART 2020: Enabling the low carbon economy in the information age report by The Climate Group on behalf of the e-Sustainability Initiative (GeSI), 2008.

[HER 12] HERZOG C., LEFEVRE L., PIERSON J.M., "Green IT for innovation and innovation for Green IT: the virtuous circle", *Human Choice and Computers (HCC10) International Conference*, Amsterdam, September 2012.

[KOO 11] KOOMEY J., BERARD S., SANCHEZ M., WONG H., "Implications of historical trends in the electrical efficiency of computing", *Annals of the History of Computing, IEEE*, vol. 33, Issue:3, p. 46-54, March 2011.

[PIE 10] PIERSON J.M., HLAVACS H., *Proceedings of the COST Action IC0804 on Energy Efficiency in Large Scale Distributed Systems*, 1st Year, IRIT, Toulouse, July 2010.

[PIE 11] PIERSON J.M., HLAVACS H., *Proceedings of the COST Action IC0804 on Energy Efficiency in Large Scale Distributed Systems*, 2nd Year, IRIT, Toulouse, July 2011.

[TUC 07] TUCKER R., "Will optical replace electronic packet switching", *SPIE Newsroom*, 2007.

[TUC 08] TUCKER R.S., BALIGA J., AYRE R., HINTON K., SORIN W.V., "Energy consumption in IP networks", *Optical Communication, ECOC 2008*, September 2008.

[ZHA 10] ZHANG YI, CHOWDHURY P., TORNATORE M., MUKHERJEE B., "Energy efficiency in telecom optical networks", *Communications Surveys & Tutorials*, IEEE, 12 (4), 2010.

Websites

[BIL] Bill Saint Arnaud. http://green-broadband.blogspot.fr.

[ECO] https://www.econet-project.eu.

[EPA] EPA, US Environmental Protection Agency ENERGY STAR Program, Report to congress on server and data center energy efficiency, available online: www.energystar.gov/ia/partners/ prod development/downloads/epa datacenter report congress final1.pdf, August 2007.

[GRE a] www.green500.org.

[GRE b] The GreenGrid, www.greengrid.org.

[GRE c] http://www.greenstarnetwork.com.

[GRE d] www.greentouch.org.

[ICO] IC0804, www.cost804.org.

[IDA]   http://www.fftelecoms.org/sites/default/files/contenus_lies/ 007.15_idate_presentation_conference_de_presse.pdf.

[PRI] http://www.efficient-datacenter.eu.

[TIC 08] Rapport TIC et Développement Durable, France, http://www.cgedd.developpement-durable.gouv.fr/IMG/pdf/0058 15-02_rapport_cle2aabb4.pdf.

[TEL]   http://www.telecomitalia.com/content/tiportal/it/innovation /events/conferences/giornata_studio_efficienzaenergeticaperchee come/jcr%3Acontent/rightParsys/linklist/linkdownloadParsys/do wnload_1/file.res/02_Cucchietti_Energia.pdf.

[TRE] www.fp7-trend.eu.

PART 1

A Step Towards Energy-efficient Networks

Chapter 2

A Step Towards Energy-efficient Wired Networks

2.1. Introduction

Whether the phenomenon stems from an increased awareness of the consequences for the environment, from a financial opportunity or from a question of reputation and business, the reduction of greenhouse gas emissions has become a primary objective in recent times. Individuals, companies and governments alike are expending a great deal of energy in reducing the energy expenditure of many sectors of activity. In parallel, information and communications technology (ICT) is increasingly present in the majority of human activities, and it is estimated that 2% of greenhouse gas emissions could be attributed to such technology, with this proportion increasing to 10% in heavily industrialized countries [GLO 07; WEB 08].

While these figures may not seem excessive at present, they will undoubtedly increase in years to come. With the

Chapter written by Aruna Prem Bianzino, Claude Chaudet, Dario Rossi and Jean-Louis Rougier.

dawn of *cloud computing*, the computation and communication infrastructures require ever-higher degrees of performance and availability. This necessitates the use of powerful hardware, which consumes a great deal of energy both because of its direct function and also of its need for cooling. In addition, the demands in terms of availability necessitate the design of superfluous setups, built on a gargantuan scale to deal with a peak load. Hence, the infrastructures are often under-used, and adapting their level of performance to the workload actually required of them is a means of optimization that appears promising on a number of levels.

The Internet, for instance, can be represented as a core network, interconnecting multiple heterogeneous access networks. These networks exhibit numerous differences in terms of technologies used, performances expected and workloads. Consequently, they offer different energy-saving opportunities. However, because of the lack of operational data and the never-ending wheel of technological advancement, it is no easy task to characterize the different sources of energy consumption and their causes, and it is impossible to reach a lasting consensus. In 2002, Roth *et al.* [ROT 02] estimated that local networks, by way of concentrators and switches, were responsible for around 80% of energy consumption by the Internet. In 2005, Nordman and Christensen [NOR 05] attributed half of the total consumption to switch matrix interface cards. In 2009, a study conducted by Deutsche Telekom [LAN 09] predicted that by 2017, the consumption of the core network would have reached the same level as that of the access networks, whereas Bolla *et al.* [BOL 11] affirm that this consumption ought to remain negligible.

From a strictly environmentalist viewpoint, the objective of green networking is to reduce the volume of greenhouse gas emissions due to the communication process. The use of renewable energy sources or of low-consumption electronics (e.g. induction devices) constitutes an obvious path for improvement. In addition, there are numerous optimization strategies related to the physical design of the infrastructure itself. For instance, it is possible to place the energy-consuming elements (data centers, etc.) close to the points of energy production so as to avoid line losses when transporting energy over long distances. It is also possible to give preference to places where the outside temperature is low all year round, thereby reducing the need for air conditioning by way of simple ventilation.

These strategies may have a significant impact on the actual energy consumption of the infrastructure; yet their influence remains marginal when it comes to the networks. For example, the delocalization of energy-consuming elements imposes constraints on the architecture of the network and alters the volume and the profile of global traffic. It is essentially a question of planning and static optimization. In this chapter, however, we shall only focus on those aspects which have a direct bearing on the dynamic function of the networks, once the design phase has been completed and the infrastructure is in place – that is, on the communication protocols. Similarly to computation infrastructures, communication networks are generally oversized and designed with a great deal of redundant capacity. Oversizing is a natural phenomenon, whereby designers can allow for changes in the volume of traffic due to new usage. In addition, because there is no management of Quality of Service (QoS), the evaluation of the traffic load at any given time is generally carried out on the basis of a measurement or an estimation of the peak traffic. As a result, during periods of low usage, the network is active but under-used, and consumes energy needlessly, even if the traffic profiles are often regular and perfectly well known. For instance, the Website *What Europeans do at Night* [WED] shows that the traffic experiences peaks during the day and troughs at night. Redundancy is necessary in order to ensure a satisfactory level of reliability and fault tolerance, but necessitates the installation of surplus machines which remain on constant alert in order to take up the baton as soon as they detect a fault. The entire issue of green networking consists of exploring possibilities for optimization while seeking to limit their impact on the QoS or fault tolerance.

In this chapter, we are only interested in forms of optimization that are applicable to a fixed infrastructure network. After presenting various models of energy consumption in section 2.2, we explore different techniques for saving energy both at the level of applications and of infrastructures in section 2.3. Then, as an example, we present a formulation of the problem of energy-efficient routing in section 2.4, before drawing our conclusions.

2.2. Models of energy consumption

Before we can outline and evaluate optimization strategies, it is essential to look at the way in which the different components of a network consume energy. While few exact figures are available for real equipment, it *is* possible to define several relevant models upon which we can construct an analysis.

It is natural, as Barroso and Hölze note [BAR 07], to imagine that the consumption of an element in a network depends on the load which is imposed upon it. Figure 2.1a shows a number of examples of energy consumption profiles of a machine varying as a function of its workload. In this figure, the metrics representing the energy and the workload have been normalized, and vary between 0 and 1. Certain machines, qualified as energy agnostic, exhibit a consumption profile which is independent of their workload, and can be turned neither on nor off. At the other extreme, we find machines whose consumption is strictly proportional to their workload. A more common model would be to consider that there are a number of modes of functioning, defining numerous levels of performance and their associated levels of consumption.

Whatever the profiles of the elements of the network, it should be noted that defining a global policy at the level of a set of resources constitutes a not-insignificant problem in terms of optimization. For instance, Figure 2.1b gives the example of two energy profiles – one optimized so as to be efficient at low demand, the other efficient with high workloads. In order to reduce the workload of certain elements, we are led to increase that of others in order to maintain the desired level of service quality. The gain made thereby must remain greater than the occasional losses incurred, and certain greedy heuristics may be inefficient when the profiles are heterogeneous.

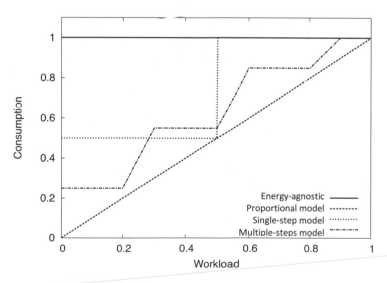

Figure 2.1a. *Examples of the evolution in the energy consumption of an element based on its workload. Energy-agnostic, proportional and staggered profiles*

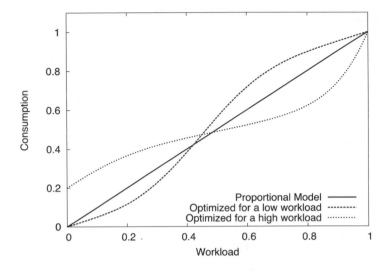

Figure 2.1b. *Examples of the evolution in the energy consumption of an element based on its workload. Profiles optimized for different modes of functioning*

Personal computers are easy to evaluate, and their architecture is similar enough to that of networking machines so that many past projects have aimed to study their energy consumption profile in detail. Some evaluations distinguish each individual component [ZHA 04; LEB 00; HYL 08] whereas others are interested in the overall, "global" consumption of a system. [RIV 08; LEW 08; RIV 08] show that there is a compromise which must be drawn between the simplicity and the precision of the models, even for the most detailed of models. Conversely, Lewis *et al.* [LEW 08] propose to use a restricted set of correlated parameters, such as the frequency of the processor, the activity of the bus or the ambient temperature, to create a linear regression model linking the energy of the system to that of its components.

The choice of software also appears to have an impact on energy consumption. Kansal and Zhao [KAN 08] offer a precise evaluation tool enabling us to conceive energy-efficient applications. The authors demonstrate the effectiveness of their method by comparing the consumption of a program operating on compressed data to that of the same program processing standard data. The compression gives rise to an additional load for the processor, but puts a lesser strain on the hard disk, resulting in better energy efficiency. It should be noted, however, that the results are very closely linked to the hardware architecture on which they are obtained; the study cited above might obtain different results were it conducted on a machine using flash memory.

As regards the interconnection equipment itself, there are few available studies, and the figures given by the manufacturers are imprecise, mentioning a single consumption value to correspond to both a particular usage profile and to peak consumption. There have been some independent evaluations published. For instance, Chabarek *et al.* [CHA 08] evaluate the consumption of two routers (Cisco 7500 and Cisco GSR12008). Hlavacs *et al.* [HLA 09] measure the consumption of four types of switch, including publicly-available and professional models, and note that the consumption is independent of the volume of traffic passing through. However, these works are few and far between, and there is a lack of representative evaluations on more complex networking equipment (DSLAMs, set-top-boxes, etc.) and comparing various technologies in terms of the principles behind them and in terms of their utilization (1 Gb/s Ethernet versus 10 Gb/s Ethernet, etc.).

At the level of a complex infrastructure such as a network, the real energy consumption is difficult to characterize, given the multitude of factors involved (redundancy, air conditioning, etc.). Baliga *et al.* [BAL 07; BAL 09] put forward a simple but pertinent Internet-type model, made up of various types of access networks (PON, FTTH, xDSL, WiMAX, etc.) and an optical core network. However, many of the hypotheses underlying that work are disputable. For instance, the consumption linked to cooling is considered to be double the nominal consumption of a given system, which is undoubtedly not the case in general. Furthermore, the impact of redundancy in the network is overlooked. Yet the authors do manage to pinpoint a lower boundary of consumption; an upper boundary has still to be added to this model.

Whether in terms of isolated machines or of networks, the community therefore lacks reliable, plentiful and universally-recognized data. For example, the evaluations based on trace measures employed in [GUP 03; GUP 04; GUN 05; GUN 06; ANA 08; PUR 06; SAB 08] use different datasets which are difficult to compare.

A number of methods have been devised to deal with this lack of data. As regards characterizing the consumption profiles, Rivoire *et al.* [RIV 07] put forward an evaluation method adapted to data centers. Mahadevan *et al.* [MAH 09] detail a method for evaluating the difference between the consumption profile of a piece of equipment and a benchmark profile perfectly proportional to the workload. They define a metric, called the Energy Proportionality Index (EPI). A complementary method which takes account of the efficiency of a machine as a function of its workload is given in [BAR 07].

While these methods can be used to compile a base of reference profiles, it is still necessary to carefully select the main metrics in order to get comparable evaluations. For instance, Ananthanarayanan and Katz [ANA 08] evaluate the total reduction in energy consumption in order to demonstrate the effectiveness of their solution, while Gupta and Singh [GUP 07] or Nedevschi *et al.* [NED 08] focus on the percentage of time spent in a state of low consumption. Bianzino *et al.* [BIA 10] list and compare the different metrics, and this overview shows that the community must reach agreement on a set of basic evaluation criteria, which are representative without being limiting, in order to reach a method of classifying solutions based on the application scenarios which make sense for them.

2.3. Energy-saving strategies

While the aforementioned evaluations may be difficult to compare, they do point to the fact that there are numerous ways in which to optimize the energy consumption of networks. Among the applications to routing and to the way in which communication links function, it is possible to act at different levels. At the planning phase of the design of a communication network, it is possible to take account of the energy-saving capability of a certain topology in addition to the usual criteria of redundancy or adaptation to the workload. At the functioning phase, the algorithms and applications can adapt to the traffic on the network, e.g. by attempting to group resources together on the same physical machines, drawing notably on the significant progress made in the field of virtualization.

2.3.1. *Transport applications and protocols*

Naturally, these applications are the root of the majority of traffic channeled through communication networks. Such applications of course define the workload which the network will have to support, but also, to a certain extent, the communication pattern. The success of peer-to-peer (P2P) applications is the undeniable proof that by seeking to improve a service – in this case, file transfer – we can also influence the profile of traffic in the networks. The use of *content delivery networks* or *proxies* in the interests of security and anonymity are also related to the application layer of the OSI model.

Optimizing the behavior of applications may therefore have a significant influence on the energy consumption of networks in order to channel their traffic. From a general point of view, Kansal and Zhao [KAN 08] and Baek and Chilimbi [BAE 09] put forward methods for profiling the energy consumption of applications and for developing energy-efficient applications. Many other contributions to the debate have been aimed at optimizing the function of specific applications.

Blackburn and Christensen [BLA 08] look at a reworking of the Telnet protocol with a view to energy saving. The modification enables the client machine to enter sleep mode and facilitates the subsequent resumption, and relies on explicit signaling so as to avoid data losses by "timeout", without the need for keep-alive messages.

In Green BitTorrent [BLA 09], the participants in a peer-to-peer file exchange network give preference to active peers for the download of parts of files, and only use inactive peers when necessary. A survey mechanism is put in place to test peers whose status is unknown − e.g. those which are announced by the tracker. However, this article does not deal with how to maintain an up-to-date view of the status of the different peer machines without causing a great deal of signaling traffic; nor does it mention how to "wake up" peer machines in sleep mode quickly, given that the Wake-on-LAN it does mention, poses security risks. The integration of these functions in set-top boxes could represent an interesting avenue for improvement.

However, these proposals remain specific. While the design of applications which favor the entry of the terminals into sleep mode is an interesting direction, taking action at the level of protocol stacks could yield significant energy savings more directly and more efficiently, given that these optimizations would be shared by multiple applications. For example, Wang and Singh [WAN 04] analyze the energy consumption due to the algorithmic complexity of TCP on different operating systems. The authors estimate the consumption due to TCP alone to be 15% of the total, of which between a fifth and a third is attributable to just the computation of the Checksum function. Irish and Christensen [IRI 98] propose to introduce explicit signaling at the level of the transport layer, by way of an option in the TCP header (TCP_SLEEP). When such a signal is received, the machine in question would queue the packets generated rather than sending them immediately. The implementation of such a mechanism requires specifying a number of details, such as the maximum frequency of sleep mode entry: parameters which are not evaluated in [IRI 98].

2.3.1.1. *Virtualization, migration and delegation of services*

Many applications require little or no interaction with a user. The services associated with these applications can therefore be delegated to particular components or moved at will within a network, so long as their level of performance remains acceptable.

For instance, Gunaratne *et al.* [GUN 05] and Nedevschi *et al.* [NED 09] show that the majority of the volume of traffic received by the interface of a personal computer can simply be ignored when the interface or the terminal is in sleep mode. The traffic related to the announcement of services or the discovery of ports requires only a minimum of computation, *a priori*. Similarly, ARP traffic management, ICMP echo requests or confirmations related to DHCP leases may be delegated to the processor of the network card. Purushothaman *et al.* [PUR 06] put forward a solution which enables a terminal to enter sleep mode without losing its network connection. However, they do not evaluate the effect of this on the machine's wake-up time; given that the machine still appears as present on the network, it may receive a request at any time. In [SAB 08], the authors evaluate the performance of such a delegation strategy, classifying the traffic experienced by standard hardware. The solution proposed supports a data rate of up to 1 Gb/s in its software version (on Smart-NIC) and up to 10 Gb/s in its hardware version, which consumes 75% less energy than its software version. Finally, going a little further with this logic, Agarwal *et al.* [AGA 09] propose to entrust the switch matrix interface with certain tasks which are routinely carried out by a machine's processor, such as the management of direct memory access (DMA), or network tasks which do not require the intervention of the user (FTP downloads, peer-to-peer transfers, etc.). In reality, these authors attribute the network card with a low energy microprocessor, RAM and flash memory.

Virtualization constitutes another promising technique for optimization, because it enables a set of services to be grouped together on the same physical platforms. In comparison with process migration or lightweight process migration, it enables a task and its environment to be migrated, reducing the complexity of this operation and delivering resistance to issues of heterogeneity and context synchronization. If one machine functioning at a full workload consumes less power than multiple under-loaded machines, this technique may prove very effective. In data centers, this type of technique has already been successfully implemented. For instance, the United States Postal Service has virtualized 791 of its 895 physical servers [USE 07]. A great many summary articles have examined virtualization solutions from a computing angle [NAN 05] and from a network point of view [KAB 08].

Some services can be delegated to a more powerful machine on a network, particularly in the case of complex traffic (e.g. peer-to-peer activity), since such a machine can easily fulfill that role on behalf of multiple others. In a residential environment, the set-top box is an ideal candidate to perform the task, as it has resources which are available to it, and constitutes a network element which is assumed to remain on, permanently. On a more global scale, a switch matrix interface card may take charge of responding to ARP, ICMP, DHCP requests, etc. Certain articles also suggest maintaining a TCP connection for hosts in sleep mode. [GUN 05; PUR 06; JIM 07] evaluated this type of delegation in the context of peer-to-peer traffic, and showed that it was possible to make substantial savings without disrupting the running of the system. Nedevschi *et al.* [NED 09] compare four types of delegation, with different degrees of complexity and different implementations, such as Click modular routers. They show that while the energy gain is significant, the trivial strategies are insufficient, particularly when there is abundant unicast traffic.

These migration and delegation techniques enable us to noticeably influence the behavior of a network. It is possible to concentrate services on a small number of machines and to put part of the network on standby, just as it is possible to control the data rate emitted and received by the elements of the network so as to keep them below a certain threshold, giving preference to the function of low-load communications links and equipment in order to take advantage of potentially proportional or near-proportional modes of consumption.

2.3.2. *Communications links*

Various empirical measurements [CHA 08; HLA 09; MAH 09] have shown that the energy consumption of an Ethernet link is independent of its actual usage. For example, on high-speed Ethernet connections (100 Mb/s and over), the link remains permanently active. The interface cards in effect maintain constant synchronization so as not to have to undergo resynchronization for every frame transmission. Consequently, the energy consumption of such a link depends only on the agreed data rate and not on the actual workload. The resultant consumption profile exhibits a staggered, "stairwell" shape. However, in this context, two approaches are possible: put certain links of the network on standby while maintaining the Quality of Service, or renegotiate the data rate based on the workload.

The IEEE 802.3az standard [IEE b] (*Energy Efficient Ethernet*) was ratified in September 2010. A history and an evaluation of this standard are given in [CHR 10]. IEEE 802.3az defines a set of signaling messages, classified under the umbrella term *Low Power Idle* (LPI), to cause links to stand-by when they are inactive, and consequently provides a fundamental tool to control the consumption of communication links. However, as many previous articles have shown [GUP 03; GUP 04; GUP 07], determining and implementing the correct compromise between reactivity and energy efficiency is no easy task. Gupta and Singh [GUP 03] propose that the nodes themselves could manage their own standby periods, based on the time between packet arrivals.

However, it must be noted that the effectiveness of a management strategy is highly dependent on the behavior of the interface card, and therefore of the technology, when entering sleep mode. An interface may indeed be in hibernation mode (deep sleep), in which case, any incoming packets will be ignored. It can use a buffer to store the packets until it is able to process them. Alternatively, it may be completely awoken whenever a packet arrives – an active standby, which gives rise to a small amount of consumption, and non-null latency. Finally, in certain cases such as with parallel machines, it is possible to use a shadow port to process the data packets instead of the inactive interfaces [ANA 08].

Gupta *et al.* [GUP 04] model the process of an interface entering standby mode as a machine with two states: fully active or standby mode. Each transition necessitates a certain amount of time – the wakeup time, which is estimated to be around 0.1 ms in modern technology – and engenders a peak in consumption. The reverse operation (standby) is assumed to be instantaneous and to have no cost in terms of energy. This model, which is simple but not simplistic, can easily be extended – e.g. to take account of the multiple transmission rates offered by numerous forms of technology, for which the energy consumption profiles differ.

Ethernet, for instance, is currently capable of offering data rates between 10 Mb/s and 10 Gb/s. The authors of [GUN 05] show that the difference in energy consumption between these data rates is not insignificant. On a typical personal computer, going from 10 Mb/s to 1 Gb/s introduces an additional consumption of 3 W, which, in 2005, represented 5% of the total consumption. For interface cards, the same alteration of the data flow engenders an extra consumption of 1.5 W per interface. The problem of selecting a data rate within a limited range of possibilities may be expressed as a problem of multicommodity flow in integers, the objective of which is to minimize the overall consumption whilst still preserving the QoS; this is known to be an NP-hard problem [EVE 75]. Various authors propose a set of strategies for adaptation of the data rates based either on a measurement of the instantaneous state of the system [GUN 06] or on its history [GUN 08].

Nedevschi *et al.* [NED 08], for a complete infrastructure, compare the strategies for standby and for selection of link rates in terms of end-to-end delay, extent of losses caused and energy saved. The energy gain is measured by the number of machines which can be deactivated in the case of standby, and by the average reduction in link rates in the case of data rate selection. Unsurprisingly, the article shows that there is a usage threshold below which it is more effective to place parts of the system on standby than to adapt the data rate of the links. Other work, such as that of

Meisner *et al.* [MEI 09] or Wierman *et al.* [WIE 09], compares standby and data rate adaptation modes for processors and servers, respectively. When standby strategies are not enormously complex, they yield better performances if we wish simply to minimize the energy consumed and the transition time. However, data rate adaptation strategies are less prone to surges and errors in measurement.

If we wish to go further than this, the definition of the conditions that trigger the switch from standby mode to active mode and vice versa may be rather tricky. Gunaratne *et al.* [GUN 05] suggest basing this decision on the state of the queues of data, and define two threshold values to trigger these transitions. Thus use of two thresholds means that the probability of rapid oscillations between the two modes can be attenuated, although it does not prevent such oscillations entirely. Gunaratne *et al.* [GUN 08] indeed show that when the data rate channeled by a link approaches its maximum capacity, this type of oscillation becomes more frequent. Consequently, they suggest coupling this mechanism with a measurement of the time spent in each state in order to define the values of the thresholds dynamically. Ananthanarayanan and Katz [ANA 08] propose to spark these state changes based on a measurement of the state of the queues over a period of time rather than on an instantaneous measurement. The Global Action Plan [GUP 07] is based on a prediction of the future state of these queues based on the current state and the characteristics of the process of data arrival.

Finally, the synchronization of the different pieces of equipment also poses a number of practical problems. When an interface decides to switch mode, it has to announce this decision to its counterpart on the other end of the link. In [GUP 03], an interface informs its neighbors just before going into standby mode, and sends a wake-up packet to its "sleeping" neighbors when it needs to transmit a frame. Gunaratne *et al.* [GUN 06; GUN 08] point out that Ethernet's procedure for self-negotiation of the data rate is too slow to facilitate a dynamic adaption of the data rate on the links. The announced latency is approximately 256 ms for a data rate of 1 Gb/s. As a result, they propose a more rapid exchange based on control packets at the MAC level, capable of completing a rate negotiation in about 100 µs at a data rate of 1 Gb/s.

2.3.2.1. *A step towards energy-efficient networks*

When we wish to optimize the operation of an entire network, it is of course possible to take measures during the phases of design or installation of the network. The use of optical hardware such as DWDM networks is considered to be energy efficient, as well as offering very high capacity. However, these technologies are fairly rigid, particularly due to the constraints related to the electronics. It is impossible to create a buffer while remaining in the optical domain, given that this mechanism is at the core of optical burst switching [QIA 99; JUE 05], which limits the capabilities for analysis and processing of packets.

Optimizing the function of an operational network, and thereby making the transition from a local optimization technique to an efficient global strategy is a difficult problem which necessitates at least a certain degree of coordination between the different elements. Chabarek *et al.* [CHA 08], and Sansò and Mellah [SAN 09] formalize this scenario as an optimization problem. Chabarek *et al.* [CHA 08] introduces the energy cost into a problem of multicommodity flow and look at the compromise between performance and energy. Sansò and Mellah [SAN 09] add a similar evaluation of the aspects to fault tolerance.

Nedevschi *et al.* [NED 08] discuss the problem of adapting the data rates of links in a complete infrastructure. The traffic is adapted upon entering the network, the packets destined for the same output router being grouped into a single burst, akin to bursts in optical switching [QIA 99; JUE 05]. This approach increases the end-to-end delay when data comes into the network, but its effect remains limited because it favors a good alternation between periods of sleep and activity of the core equipment. The authors examine the influence of the time spent on standby on the load of the network, the size of the bursts and the transition time. They show that the added complexity is limited, but they do not offer recommendations about the duration or management of periods of inactivity.

Finally, at the level of routing, we can, when the load allows, seek to aggregate the data streams on a reduced set of machines and network connections, thus favoring the standby of other machines. This possibility was mentioned in the position paper [GUP 03] as a possible avenue for optimization. This article adopts the example of two parallel routers located on the boundary of an autonomous system. The evolution of the routing protocol is then mentioned as a prerequisite for coordinating the sleep periods of those two routers. OSPF considers the links on standby as a link failure and updates the path topology, triggering a process that is too unwieldy to be executed frequently. IBGP suffers oscillations between the paths, and no shortage of occasional loops. The article argues for alternation between a number of pre-calculated solutions, which relies on the use of a centralized decision point.

This routing, which can be implemented by way of a dynamic configuration of the weights of the links, must ensure the connectivity of the network is preserved, and that it has no notable impact on the QoS. Consequently, it is necessary to ensure a certain level of diversity of paths and limit the maximum data rate channeled through each link in order to safeguard its efficiency. Formally, this is also a multicommodity flow problem [CHI 09] which is expressed as an integer linear programming problem. The article evaluates some greedy heuristics which consist of switching off certain links and nodes. Since the authors study the case of a simple provider equipped with multiple boundary links, the scenario can be considered a best-case scenario, because the redundancy improves the efficiency of the solution. Other studies choose to determine the optimal point by solving the problem numerically but only look at the links on the network [FIS 10].

2.4. The problem of energy-efficient routing

As indicated before, energy-efficient routing generally seeks to aggregate data flows on a subset of machines and network links in order to allow other resources to enter sleep mode. As an example, here we present a study based on a model which considers the links and nodes of a network, going further than the model used by [FIS 10]. The problem of routing is expressed in the form of a problem of optimization, which we solve numerically for several models of energy consumption, implemented on a real topology with a real traffic matrix.

2.4.1. *Model of energy consumption*

As indicated in section 2.2, it is difficult for a model to be based on real data regarding energy consumption. Therefore, we choose here to base our work on a generic and easily extensible model of consumption, for which the values will be borrowed from various publications [GUN 05; GUN 06; TUC 08; HWM 08]. Despite numerous differences, it is commonly accepted that the energy consumption of an interconnecting element may be approximated by a refined increasing function between a minimal value, E_0, which represents a state of passivity, and a maximal value, M

[BAR 07]. While many elements have a consumption profile in step form, the refined model remains a good approximation because the real function is increasing, and in general exhibits a fairly constant gradient. In addition, we shall consider that the consumption is zero when the element is not in use. This model, which we term "*idleEnergy*", is represented in Figure 2.2 by a solid line.

For the values of the parameters E_0 and M, we base our reasoning on the figures which are most commonly found in the existing body of literature. Table 2.1 gives an overview of these parameters. C represents the switching capacity of a node. Since the maximum switching capacity of a node is a figure which is absent from the literature, we shall consider that a switch matrix is capable of managing half the sum of the capacities of the links connected to it. This somewhat conservative choice of value enables us to model a powerful and adaptable switch-matrix.

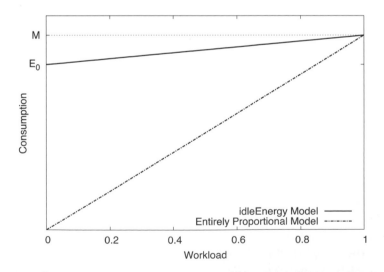

Figure 2.2. *Parametric models of the energy consumption of a switch matrix as a function of its workload*

Network element	E_0 [Watts]	M [Watts]	References
Nodes	$0.85 \; C^{3/2}$	$C^{3/2}$	[TUC 08]
Link of capacity (0-100) Mb/s	0.48	0.48	[HWM 08, GUN 05]
Link of capacity (100-600) Mb/s	0.9	1.00	[HWM 08, GUN 05]
Link of capacity (600-1,000) Mb/s	1.7	2.00	[GUN 06]

Table 2.1. *Parametric energy consumption (in Watts) of various interconnecting elements*

Two particular cases of this model are particularly pertinent in our analysis and will give us boundaries for our study. In the first case, represented in Figure 2.2 by a dotted line and which we shall call the entirely proportional model, the parameter E_0 is equal to 0 and the consumption increases in a linear fashion with increasing usage. This model is pertinent in the case of fully adaptive elements and represents the case of links conforming to the data rate adaptation model [GUN 08]. Conversely, in the energy-agnostic model, represented in Figure 2.2 by the dotted line along the top of the figure, the energy consumption is independent of the usage and the elements are either active or turned off.

2.4.2. *Formulation of the problem*

We represent a network by a directed graph, $G = (N, L)$, where N is the set of vertices, indiscriminately representing the sources and destinations of the traffic as well as the interconnecting elements, L is the set of arcs representing the communication links. For every component of the network a (be it a node or a link), we shall use the notation l_a for its load and c_a for its capacity, i.e. the maximum load which it can handle.

Our goal is to determine the configuration of the network – i.e. the power state and the load of the different nodes and links – which minimizes the overall energy consumed. This consumption is expressed as the sum of the individual consumptions of the links and the nodes. The consumption of each element is modeled, as previously stated, as a refined function of its load. We use a binary variable, x_a, to denote the status (on or off) of each element a ($x_a = 1$ when a is on and $x_a = 0$ otherwise). The gradient of the function characterizing the element a is denoted E_{fa}. Finally, we shall consider that the links are bidirectional and that they are fully switched on as soon as a request is detected in either direction. Since we are using a directed graph, the load of a link is the sum of its loads in both directions. The total energy consumed is then expressed in accordance with equation [2.1], wherein the first component must be divided by two in order to avoid counting each link twice:

$$\frac{1}{2} \sum_{(i,j)\in L} \left(\frac{(l_{ij} + l_{ji})E_{fij}}{c_{ij}} + x_{ij}E_{0ij} \right) + \sum_{n\in N} \left(\frac{l_n E_{fn}}{c_n} + x_n E_{0n} \right) \qquad [2.1]$$

The load imposed on the network is defined by a traffic matrix which, for each couple (s, d) of input and output nodes, indicates the volume of traffic flowing from s to d. This volume is represented by r_{sd} hereafter. This flow is routed into the network, causing traffic on the link (i, j) chosen. This traffic matrix defines a set of constraints as follows:

$$\sum_{(i,s,d)\in N^3} f_{ij}^{sd} - \sum_{(i,s,d)\in N^3} f_{ji}^{sd} = \begin{cases} r_{sd} & \forall(s,d) \in N^2, j = s \\ -r_{sd} & \forall(s,d) \in N^2, j = d \\ 0 & \forall(s,d) \in N^2, j \neq s,d \end{cases} \qquad [2.2]$$

In order to preserve the QoS, the load of the links should never reach 100%, but should remain below an arbitrary value α which the administrator deems reasonable. This constraint is expressed as follows:

$$\sum_{(s,d)\in N^2} f_{ij}^{sd} = l_{ij} \leq \alpha c_{ij} \quad \forall (i,j) \in L \qquad\qquad [2.3]$$

Hereafter, we assume that the load of a node is directly proportional to the traffic entering and exiting the node. Since we are only looking at the interconnecting elements (the sources and destinations of the traffic being the input or output routers), we shall consider hereafter that these two values are identical, which results in the following set of constraints:

$$l_n = \sum_{(i,n)\in L} l_{in} + \sum_{(n,i)\in L} l_{ni} \quad \forall n \in N \qquad\qquad [2.4]$$

Finally, we suppose that a node or a link is switched off when its load reaches 0, which links the variables x_a and l_a in the following manner for any element of the network:

$$Z x_{ij} \geq l_{ij} + l_{ji} \quad \forall i,j \in L \qquad\qquad [2.5]$$

$$Z x_n \geq l_n \quad \forall n \in N \qquad\qquad [2.6]$$

where Z is a "large" number (meaning at least double the maximum between the capacities of the nodes and links), used to force the variable x_a to assume a value of 1 when the load of a is greater than 0, and a value of 0 when $l_a = 0$.

Energy-efficient routing therefore seeks to minimize the energy consumption defined by equation [2.1], while respecting this set of constraints. The problem in question is a mixed integer linear programming problem with binary variables (x_a) and real variables (l_a).

2.4.3. Experimental results

Just as it is difficult to find precise figures on energy consumption, there is no consensus on one or more representative and pertinent scenarios. However, in the absence of such scenarios, it is easy to find a case where the energy gain from a certain algorithm is extraordinary, in spite of the realism. We have chosen instead to base our study on a realistic scenario particularly unfavorable to the approach in question in order to identify a lower boundary of the potential gain.

We have chosen to use the GEANT network topology [GEA], shown in Figure 2.3. This real and reasonably-complex network is composed of 23 nodes and 74 links. We consider 24 traffic matrices: one per hour between 0:30 and 23:30 on a standard weekday, out of the available matrices. The routing on this network is assumed to be defined by the IGP Weight-Optimization (IGP-WO) algorithm [IGP], which is the standard for network operators. Hereafter, this scenario will be called IGP-WO routing.

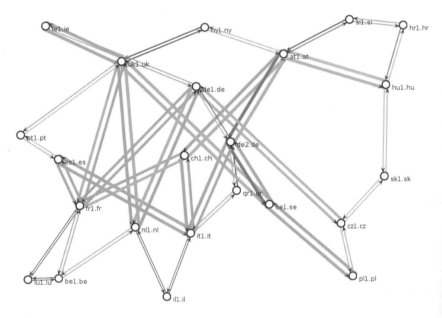

Figure 2.3. *The GEANT topology; the different shades of gray of the links represent their level of usage*

As regards the *idleEnergy* model, the energy gain can largely be attributed to the shutting down of certain elements of the network, because it enables the constant weight E_0 to be reduced. It is clear, considering the values given in Table 2.1, that the impact of this constant factor is of greater importance than the gain resulting from a load adaptation ($M - E_0$). Also, in our model, the energy consumption of the links is lower than that of the nodes by one order of magnitude, which means that the potential gain relating to the links is slight. However, in the topology under discussion here, it is impossible to switch a node off, because every single node is both the source and destination of a non-null traffic flow. In this sense, the GEANT scenario represents a case which is deeply unfavorable for this strategy.

The corresponding problem of optimization has been modeled using AMPL [AMP] and numerically solved by CPLEX [IBM]. The average results obtained on the 24 traffic matrices are summed up in Tables 2.2 and 2.3 for the three energy models, while Figure 2.4 illustrates the energy gain, separating the contributions of the links and the nodes.

Scenario	IGP-WO routing		
	Nodes	Links	Total
Energy-agnostic	7,676.00	59.12	7,735.12
idleEnergy	6,565.95	46.23	6,612.18
Entirely proportional	307.21	10.97	318.18

Table 2.2. *IGP-WO energy consumption, for different models of consumption (average values for the 24 traffic matrices)*

Scenario	Green routing					
	Nodes		**Links**		**Total**	
Energy-agnostic	7,676.00	(-0.0)	59.12	(-0.0)	7,735.12	(-0.0)
idleEnergy	6,569.22	(+0.05)	30.34	(-34.4)	6,599.56	(-0.2)
Entirely proportional	286.69	(-6.7)	5.10	(-53.5)	291.79	(-8.3)

Table 2.3. *Energy consumption with green routing, for different models of consumption (average values for the 24 traffic matrices); the numbers in parentheses represent the gain in comparison to the IGP-WO case shown above*

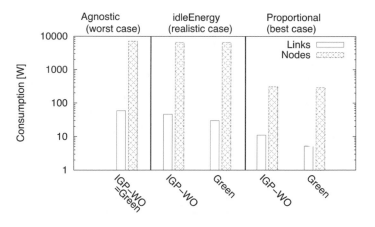

Figure 2.4. *Energy consumption of different routing algorithms for different energy models*

The results cited here confirm the hypotheses mentioned above. Little gain can be attributable to the links, which accounts for the modest gain in energy yielded by the *idleEnergy* model. Figure 2.5 details the variations in consumption for this model throughout the day, for the IGP-WO and green routing strategies. We can see that green routing leads to a slight energy saving, but one which increases as the load on the network becomes greater.

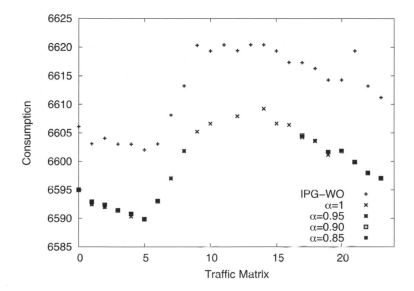

Figure 2.5. *Comparison of energy consumption by the routing algorithms throughout the day (idleEnergy model)*

If we now examine the entirely proportional model of consumption, the energy gains are due to the aggregation of the traffic on those paths containing the most energy-efficient elements. In this model, we are not seeking to turn off the nodes or the links, because the energy consumption in passive mode is null ($E_0 = 0$). The results shown in Table 2.2 show that it is possible, in this case, to obtain a far greater gain with this type of consumption profile. This demonstrates the advantage of software and hardware bricks approaching the consumption profiles of a proportional model such as the adaptation of the data rate on the link proposed by IEEE 802.3az [IEEa] or the dynamic adaptation of voltage and frequency in electronics [ISC 06].

2.4.3.1. *Impact on Quality of Service*

In the green strategy, the gains in energy are due to the shutdown of elements of the network or to the optimization of their load so as to reduce their consumption. This strategy runs counter to the conventional practice of redundancy to improve the fault tolerance and load balancing. Hence it is necessary to evaluate the impact of this strategy on the load of the links and compare it with the case of the IGP-WO standard.

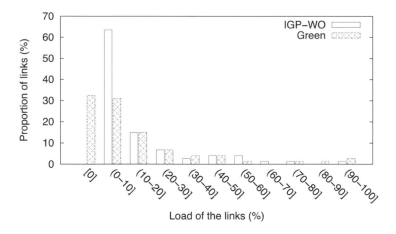

Figure 2.6. *Distribution of the load of the links for the IGP-WO and green routing schemes*

More specifically, our interest lies in characterizing the way in which a green solution shifts the traffic, and the impact that routing has on the load of the elements, since that performance indicator has a direct bearing on the QoS. For simplicity's sake, here we present only the results relating to the scenario corresponding to the time of day 0:30 and the *idleEnergy* model. Indeed, the conclusions are similar for the other traffic matrices. Figure 2.6 represents the distribution of the loads of the links for the two routing schemes being compared. It should be noted that in the case of IGP-WO, no links are inactive, whereas the green routing approach enables many of these links to be deactivated. Consequently, the number of links with a high workload is naturally greater with green routing.

Figure 2.7 shows the average load of the links in both scenarios. Firstly, we can see a slight increase in the average load in the case of green routing, because of a slight increase in the length of the paths. Secondly, green routing tends to shift the load from links with mediocre capacity to high-capacity links. We also note that the aggregation of the traffic is often impossible for a certain number of low-capacity access links, given that these areas of the network are generally more constrained and the diversity is lesser.

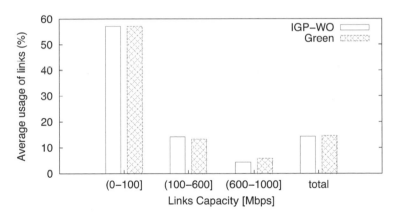

Figure 2.7. *Average load of the links by type*

Figure 2.8 identifies the links which are actually switched off by the green strategy. These links are represented by heavy black lines. In this scenario, the procedure for shutting down the links only selects those links which are not heavily loaded (a load of around 5.2% on average in the agnostic scenario) and those of high capacity (all the links whose capacity is lower than 100 Mbps remain active). Overall, the nodes connected to such a link are connected to another similar link.

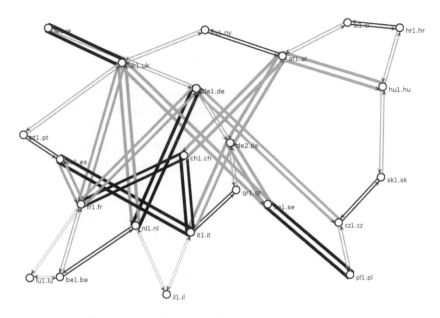

Figure 2.8. *Elements of the network deactivated*
by green routing (heavy black lines)

These various observations, even though they are only made on one specific scenario, lead us to think that the impact of green routing on Quality of Service is reasonable, although not non-existent. The traffic is redirected on high-capacity links without having any significant impact on their load. In addition, it is possible in this model to limit the maximum usage of the links by way of the parameter α described above. Figure 2.9 demonstrates the influence of this parameter. The results in this figure represent the average obtained on the 24 traffic matrices using the entirely proportional model. The conclusions are similar for the other models of consumption. This figure shows that the total energy gain is not greatly affected by the introduction of this limitation. By design, the nodes are never loaded to more than 50%. However, the effect of this reduction on the links is more noticeable: in many cases, the problem becomes unsolvable. Indeed, in the low-load scenario (traffic matrix at

0:30), certain links are loaded to more than 90% with IGP-WO. Consequently, limiting the maximum load on the links does not yield any optimization. The percentage of achievable solutions is represented on the right-hand axis in Figure 2.9.

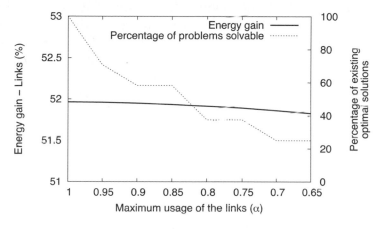

Figure 2.9. *Percentage energy saving as a function of the maximum load permitted on the links*

2.4.3.2. *Sensitivity*

In order to study the sensitivity of the green routing scheme and evaluate its potential for less restricted topologies, here we present the results obtained for a number of modifications to the scenario in question. We consider that one or more nodes from the GEANT network no longer send or receive traffic but instead become solely core nodes, dedicated to the task of interconnection. These core nodes are chosen from the five most central nodes, *at1.at*, *ch1.ch*, *de1.de*, *es1.es* and *uk1.uk*. We tested the set of possible configurations for 1, 2..., 5 core nodes.

The results shown in Figure 2.10 for the *idleEnergy* model on the traffic matrix at 05:30 correspond to the minimum load on the network, but not to the maximum energy gain, as Figure 2.5 shows.

The possibility of switching nodes off should give rise to a significant potential for optimization. For $N = 1$, the total gain in the *idleEnergy* model is approximately 6% – which corresponds to 30 times the 0.2% gain obtained by only switching links off. This 6% corresponds to the shutdown of one twenty-third of the nodes. As Figure 2.10 shows, this tendency is not confirmed when we increase the number of core nodes, because it is not always possible to turn these nodes off in view of the routing constraints. The curve formed by the squares in Figure 2.10 represents, on the right-hand axis, the average number of nodes actually switched off, which increases linearly with the number of core nodes. The "green" curve represents the lower limit of consumption obtained by switching off all the core nodes.

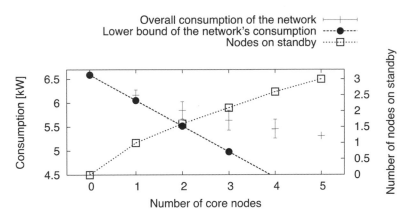

Figure 2.10. *Energy savings as a function of the number of nodes in the core*

This analysis shows that even if redundancy is high in the network, green routing does in fact enable us to aggregate the traffic on certain links, and therefore facilitates the design of a topology to handle peak traffic. However, the actual energy gain to be expected from green routing is highly dependent upon the distribution of traffic and routing.

2.5. Conclusion

In this chapter, we have presented a number of possibilities for optimizing the energy consumption of a wired communication network and underlined the lack of precise and realistic models in support of the design of protocols and algorithms. While various traffic profiles can be envisaged, the real values of consumption and adaptation of the load of this consumption are too often absent. In particular, the model of consumption of the communication links remains imprecise. Whereas strategies such as IEEE 802.3az tend to make Ethernet links behave closer to a proportional model, the consumption of the optical links, for instance, is more dependent on the distance to be covered than on the load in terms of data. Radio networks, which are not discussed in this chapter, constitute an undeniable hotbed for optimization because the processing of a radiofrequency signal is an operation which is very costly in terms of energy.

In the second part of the chapter, we studied the example of energy-efficient routing, whose conclusions are limited if we wish to preserve the Quality of Service of the network. The effectiveness of this type of mechanism is heavily dependent both on the profile of the traffic which the network has to channel, but also on the mode of energy consumption of the different pieces of equipment and on their capacity to adapt their activity to the load imposed upon them. In the case of a binary (on or off) profile of consumption, all the potential for optimization lies in the routing of the traffic and is therefore dependent on the redundancy in the network. While we can expect to benefit from such redundancy in general, since networks are designed to be resistant to certain breakdowns, there is nothing to guarantee that the traffic profile will not greatly limit the potential for optimization, unless this objective is taken into account when a network is being designed.

Finally, the goal of our work here is to simply characterize the potential of green strategies on the basis of real-life scenarios. To define an algorithm to assign weight in an operational network based on the load is also a complex task, and it would be interesting to contrast the various heuristics proposed in the existing literature such as [CHI 09] against these scenarios and compare them to the optimal strategy.

2.6. Bibliography

[AGA 09] AGARWAL Y., HODGES S., CHANDRA R., SCOTT J., BAHL P., GUPTA R., "Somniloquy: augmenting network interfaces to reduce PC energy usage", *Proceedings of the 6th USENIX Symposium on Networked Systems Design and Implementation (NSDI)*, Boston, Massachusetts, United States, April 2009.

[ANA 08] ANANTHANARAYANAN G., KATZ R.H., "Greening the switch", *Proceedings of the USENIX Workshop on Power Aware Computing and Systems (HotPower), held at the Symposium on Operating Systems Design and Implementation (OSDI 2008)*, San Diego, California, United States, December 2008.

[BAE 09] BAEK W., CHILIMBI T., Green: a system for supporting energy-conscious programming using principled approximation, report no. MSR-TR-2009-89, Microsoft Research, July 2009.

[BAL 07] BALIGA J., HINTON K., TUCKER R.S., "Energy consumption of the Internet", *Proceedings of the Joint International Conference on Optical Internet and the 32nd Australian Conference on Optical Fibre Technology (COIN-ACOFT 2007)*, p. 1-3, Melbourne, Australia, June 2007.

[BAL 09] BALIGA J., AYRE R., SORIN W.V., HINTON K., TUCKER R.S., "Energy consumption in optical IP networks", *Journal of Lightwave Technology*, vol. 27, no. 13, p. 2391-2403, 2009.

[BAR 07] BARROSO L.A., HÖLZLE U., "The case for energy-proportional computing", *IEEE Computer*, vol. 40, no. 12, p. 33-37, 2007.

[BIA 10] BIANZINO A.P., RAJU A., ROSSI D., "Apple-to-Apple: a common framework for energy-efficiency in networks", *Proceedings of ACM SIGMETRICS, GreenMetrics workshop*, New York, United States, June 2010.

[BLA 08] BLACKBURN J., CHRISTENSEN K., "Green Telnet: modifying a client-server application to save energy", *Dr. Dobb's Journal*, October 2008.

[BLA 09] BLACKBURN J., CHRISTENSEN K., "A simulation study of a new green BitTorrent", *Proceedings of the First International Workshop on Green Communications (GreenComm) in conjunction with the IEEE International Conference on Communications*, Dresden, Germany, June 2009.

[BOL 11] BOLLA R., BRUSCHI R., CHRISTENSEN K., CUCCHIETTI F., DAVOLI F., SINGH S., "The potential impact of green technologies in next generation wireline networks. Is there room for energy savings optimization?", *IEEE Communication Magazine*, vol. 49, no. 8, 2011.

[CHA 08] CHABAREK J., SOMMERS J., BARFORD P., ESTAN C., TSIANG D., WRIGHT S., "Power awareness in network design and routing", *Proceedings of the 27th Annual Conference on Computer Communications (IEEE INFOCOM 2008)*, p. 457-465, Phoenix, Arizona, United States, April 2008.

[CHI 09] CHIARAVIGLIO L., MELLIA M., NERI F., "Reducing power consumption in backbone networks", *Proceedings of the IEEE International Conference on Communications (ICC 2009)*, Dresden, Germany, June 2009.

[CHR 10] CHRISTENSEN K., REVIRIEGO P., NORDMAN B., BENNETT M., MOSTOWFI M., MAESTRO J.A., "IEEE 802.3az: the road to energy efficient ethernet", *IEEE Communication Magazine*, vol. 48, no. 11, 2010.

[EVE 75] EVEN S., ITAI A., SHAMIR A., "On the complexity of time table and multi-commodity flow problems", *Proceedings of the 16th Annual Symposium on Foundations of Computer Science (SFCS'75), IEEE Computer Society*, p. 184-193, Washington, United States, October 1975.

[FIS 10] FISHER W., SUCHARA M., REXFORD J., "Greening backbone networks: reducing energy consumption by shutting off cables in bundled links", *Proceedings of 1st ACM SIGCOMM Workshop on Green Networking*, New Delhi, India, August 2010.

[GLO 07] GLOBAL ACTION PLAN, An inefficient truth, Global Action Plan Report, http://globalactionplan.org.uk, December 2007.

[GUN 05] GUNARATNE C., CHRISTENSEN K., NORDMAN B., "Managing energy consumption costs in desktop PCs and LAN switches with proxying, split TCP connections and scaling of link speed", *International Journal of Network Management*, vol. 15, no. 5, p. 297-310, 2005.

[GUN 06] GUNARATNE C., CHRISTENSEN K., SUEN S.W., "Ethernet adaptive link rate (ALR): analysis of a buffer threshold policy", *Proceedings of the IEEE Global Communications Conference (GLOBECOM 2006)*, San Francisco, California, United States, November 2006.

[GUN 08] GUNARATNE C., CHRISTENSEN K., NORDMAN B., SUEN S., "Reducing the energy consumption of ethernet with adaptive link rate (ALR)", *IEEE Transactions on Computers*, vol. 57, no. 4, p. 448-461, 2008.

[GUP 03] GUPTA M., SINGH S., "Greening of the Internet", *Proceedings of the ACM Conference on Applications, Technologies, Architectures, and Protocols for Computer Communications (SIGCOMM 2003)*, p. 19-26, Karlsruhe, Germany, August 2003.

[GUP 04] GUPTA M., GROVER S., SINGH S., "A feasibility study for power management in LAN switches", *Proceedings of the 12th IEEE International Conference on Network Protocols (ICNP 2004)*, p. 361-371, Berlin, Germany, October 2004.

[GUP 07] GUPTA M., SINGH S., "Using low-power modes for energy conservation in ethernet LANs", *Proceedings of the 26th Annual IEEE Conference on Computer Communications (IEEE INFOCOM 2007)*, p. 2451-2455, Anchorage, Alaska, May 2007.

[HAY 08] HAYS R., WERTHEIMER A., MANN E., "Active/idle toggling with low-power idle", *Presentation for IEEE 802.3az Task Force Group Meeting*, January 2008.

[HLA 09] HLAVACS H., DA COSTA G., PIERSON J.M., "Energy consumption of residential and professional switches", *Proceedings of the IEEE International Conference on Computational Science and Engineering*, IEEE Computer Society, vol. 1, p. 240-246, 2009.

[HYL 08] HYLICK A., SOHAN R., RICE A., JONES B., "An Analysis of Hard Drive Energy Consumption", *Proceedings of the IEEE International Symposium on Modeling, Analysis and Simulation of Computers and Telecommunication Systems (MASCOTS 2008)*, p. 1-10, Baltimore, Maryland, United States, September 2008.

[IRI 98] IRISH L., CHRISTENSEN K.J., "A 'Green TCP/IP' to reduce electricity consumed by computers", *Proceedings of IEEE Southeastcon'98*, Orlando, Florida, United States, April 1998.

[ISC 06] ISCI C., BUYUKTOSUNOGLU A., CHER C.Y., BOSE P., MARTONOSI M., "An analysis of efficient multi-core global power management policies: maximizing performance for a given power budget", *Proceedings of the 39th Annual IEEE/ACM International Symposium on Microarchitecture (MICRO 39)*, IEEE Computer Society, p. 347-358, Orlando, Florida, United States, December 2006.

[JIM 07] JIMENO M., CHRISTENSEN K., "A prototype power management proxy for Gnutella peer-to-peer file sharing", *Proceedings of the 32nd IEEE Conference on Local Computer Networks (LCN 2007)*, Dublin, Ireland, October 2007.

[JUE 05] JUE J.P., VOKKARANE V.M., *Optical Burst Switched Networks*, Springer, Heidelberg, Germany, 2005.

[KAB 08] KABIR CHOWDHURY N.M., BOUTABA R., A survey of network virtualization, report no. CS-2008-25, University of Waterloo, October 2008.

[KAN 08] KANSAL A., ZHAO F., "Fine-grained energy profiling for power-aware application design", *ACM SIGMETRICS Performance Evaluation Review*, vol. 36, no. 2, p. 26-31, 2008.

[LAN 09] LANGE C., "Energy-related aspects in backbone networks", *Proceedings of 35th European Conference on Optical Communication (ECOC 2009)*, Vienna, Austria, September 2009.

[LEB 00] LEBECK A.R., FAN X., ZENG H., ELLIS C., "Power aware page allocation", *ACM SIGOPS Operating Systems Review*, vol. 34, no. 5, p. 105-116, 2000.

[LEW 08] LEWIS A., GHOSH S., TZENG N.F., "Run-time energy consumption estimation based on workload in server systems", *Proceedings of the USENIX Workshop on Power Aware Computing and Systems (HotPower), held at the Symposium on Operating Systems Design and Implementation (OSDI)*, San Diego, California, United States, December 2008.

[MAH 09] MAHADEVAN P., SHARMA P., BANERJEE S., RANGANATHAN P., "A power benchmarking framework for network devices", *Proceedings of IFIP Networking 2009*, Aachen, Germany, May 2009.

[MEI 09] MEISNER D., GOLD B.T., WENISCH T.F., "PowerNap: eliminating server idle power", *Proceedings of the 14th International Conference on Architectural Support for Programming Languages and Operating Systems (ASPLOS '09)*, p. 205-216, Washington, United States, March 2009.

[NAN 05] NANDA S., CHIUEH T.C., A survey on virtualization technologies, report no. TR179, Department of Computer Science, SUNY at Stony Brook, February 2005.

[NED 08] NEDEVSCHI S., POPA L., IANNACCONE G., RATNASAMY S., WETHERALL D., "Reducing network energy consumption via sleeping and rate-adaptation", *Proceedings of the 5th USENIX Symposium on Networked Systems Design and Implementation (NDSI2008)*, San Francisco, California, United States, April 2008.

[NED 09] NEDEVSCHI S., CHANDRASHEKAR J., LIU J., NORDMAN B., RATNASAMY S., TAFT N., "Skilled in the art of being idle: reducing energy waste in networked systems", *Proceedings of the 6th USENIX Symposium on Networked Systems Design and Implementation (NSDI 2009)*, Boston, Massachusetts, United States, April 2009.

[NOR 05] NORDMAN B., CHRISTENSEN K., "Reducing the energy consumption of network devices", *IEEE 802.3 Tutorial*, July 2005.

[PUR 06] PURUSHOTHAMAN P., NAVADA M., SUBRAMANIYAN R., REARDON C., GEORGE A.D., "Power-proxying on the NIC: a case study with the Gnutella file-sharing protocol", *Proceedings of the 31st IEEE Conference on Local Computer Networks (LCN 2006)*, Tampa, Florida, United States, November 2006.

[QIA 99] QIAO C., YOO M., "Optical burst switching (OBS) – a new paradigm for an optical Internet", *Journal of High Speed Networks. Special Issue on Optical Networking*, vol. 8, no. 1, p. 69-84, IOS Press, 1999.

[RIV 07] RIVOIRE S., SHAH M.A., RANGANATHAN P., KOZYRAKIS C., "JouleSort: a balanced energy-efficiency benchmark", *Proceedings of the 2007 ACM SIGMOD International Conference on Management of Data (SIGMOD '07)*, p. 365-376, Beijing, China, June 2007.

[RIV 08] RIVOIRE S., RANGANATHAN P., KOZYRAKIS C., "A comparison of high-level full-system power models", *Proceedings of the USENIX Workshop on Power Aware Computing and Systems (HotPower), held at the Symposium on Operating Systems Design and Implementation (OSDI)*, San Diego, California, United States, December 2008.

[ROT 02] ROTH K.W., GOLDSTEIN F., KLEINMAN J., Energy consumption by office and telecommunications equipment in commercial buildings, Volume I: Energy consumption baseline, Report, National Technical Information Service (NTIS), US Department of Commerce, January 2002.

[SAB 08] SABHANATARAJAN K., GORDON-ROSS A., "A resource efficient content inspection system for next generation smart NICs", *Proceedings of the IEEE International Conference on Computer Design 2008. (ICCD 2008)*, p. 156-163, Lake Tahoe, California, United States, October 2008.

[SAN 09] SANSÒ B., MELLAH H., "On reliability, performance and internet power consumption", *Proceedings of 7th International Workshop on Design of Reliable Communication Networks (DRCN 2009)*, Washington, United States, October 2009.

[TUC 08] TUCKER R., BALIGA J., AYRE R., HINTON K., SORIN W., "Energy consumption in IP networks", *Proceedings of the 34th European Conference on Optical Communication (ECOC'08)*, Brussels, Belgium, September 2008.

[USE 07] US ENVIRONMENTAL PROTECTION AGENCY, Energy star program, report to Congress on server and data center energy efficiency public law 109-431, August 2007.

[WAN 04] WANG B., SINGH S., "Computational energy cost of TCP", *Proceedings of the 23rd Annual Joint Conference of the IEEE Computer and Communications Societies (INFOCOM 2004)*, vol.2, p. 785-795, Hong-Kong, March 2004.

[WEB 08] WEBB M., "SMART 2020: enabling the low carbon economy in the information age", *The Climate Group*, London, June 2008.

[WIE 09] WIERMAN A., ANDREW L.L.H., TANG A., "Power-aware speed scaling in processor sharing systems", *Proceedings of the 28th Annual Conference on Computer Communications (IEEE INFOCOM 2009)*, Rio de Janeiro, April 2009.

[ZHA 04] ZHAI B., BLAAUW D., SYLVESTER D., FLAUTNER K., "Theoretical and practical limits of dynamic voltage scaling", *Proceedings of the 41st Annual ACM Design Automation Conference (DAC 2004)*, p. 868-873, San Diego, California, United States, June 2004.

Websites

[AMP] AMPL, A Modeling Language For Mathematical Programming: http://www.ampl.com/.

[GEA] The Geant Network: http://www.geant.net/.

[IBM] IBM ILOG CPLEX Optimizer Homepage: http://www-01.ibm.com/software/integration/optimization/cplex-optimizer/.

[IEE a] IEEE 802.3AZ Task Force: http://www.ieee802.org/3/az/index.html.

[IEE b] IEEE P802.3AZ ENERGY Efficient ETHERNET TASK FORCE: http://www.ieee802.org/3/az/index.html.

[IGP] The Interior Gateway Protocol Weight Optimizer (Igp-Wo) Algorithm: http://totem.run.montefiore.ulg.ac.be/algos/igpwo.html.

[WED] What Europeans Do At Night: http://asert.arbornetworks.com/2009/08/what-europeans-do-at-night/.

Chapter 3

A Step Towards
Green Mobile Networks

3.1. Introduction

Like numerous sectors of the economy which are obliged to deal with the impact of their activity on global warming, the world of telecommunications – and more specifically that of mobile communications – has integrated the necessity to reduce its energy consumption into its roadmap. Thus, whether from standardization engineers or the users and operators of these networks, proposals of mechanisms and procedures which contribute to the objectives in reduction of greenhouse gas emissions have been put forward, some of which have already been implemented in the networks and systems.

Hence, we propose in this chapter to give an overview of these different techniques or technologies, through the lens of a classification depending on the angle from which each addresses the energy consumption issue.

Chapter written by Sami TABBANE.

3.1.1. *Decreasing power: an imperative in a cellular radio network*

Let us first recall the fact that cellular networks are qualified as interference-limited systems. This implies that numerous techniques used in these networks (from the transmission channel to the techniques used on radio sites) are developed to minimize the interference created by the system as a result of the frequency reuse scheme (a basic technique inherent in cellular technology). Also note that because of the power limitation of mobiles and the constraints on their energy (limited battery life), the reduction in power needed on these terminals is also a concern for the designers of cellular systems. One positive consequence of these two major issues is the reduction in the transmitted energy of the base transceiver stations or mobile terminals and therefore their consumption. We can cite power control mechanisms, DTX (Discontinuous Transmission), sleep mode (specified in WiMAX, for instance), DRX (Discontinuous Reception), handover based on the criterion of the best cell, etc.

In the context of cellular systems, power minimization has long been a factor in the concerns of the designers and operators of these networks. However, as the pressure for energy savings and reductions in greenhouse gas emissions has increased, the concept of green cellular networks has emerged.

3.1.2. *Definition of and need for green cellular*

We shall begin by defining what is commonly known as "green cellular". Green cellular comprises a set of techniques which facilitate the implementation of cellular networks which are efficient in terms of both energy and power, generating fewer greenhouse gas emissions and less radio radiation. While this concept has emerged and become prevalent in a domain which seems a long way from these issues – unlike in other domains such as transport or industry – this is mainly because the information and communication technology (ICT) industry consumes between 3 and 7% of electricity and generates around 2% of CO_2 emissions (which is equivalent to the global emissions from the aviation industry) [KEL 07; KAR 03]. The advent of green cellular is obviously the result of the growth in the telecommunications sector. Of the various actors in telecommunications networks, mobile network operators are the primary consumers of energy. For instance, in Italy, the incumbent telecommunications operator (for both landlines and mobiles) *Telecom Italia* is the second-largest consumer of energy in the country. In addition, because of the near-exponential increase of traffic on mobile networks [WIL 08], the energy consumption of these networks is growing at a faster rate than the ICT sector in general, with predictions of 6.3 exabytes per month in 2015 – equivalent to 26 times the figure for 2010 [CIS 10]. The energy consumed by information and communication technology is growing by 15-20% a year [FET 08]. Energy consumption is particularly significant for the transmission features, as 57% of the electricity consumed by a cellular network is used by its radio access network. In CDMA access technology-based 3G networks, where the base transceiver stations (BTSs or Node Bs) share their power between all mobiles at the same time, these constitute the element which consumes the most energy in the network. Thus, a 3G station, whose power is around 40 Watts, consumes around 500 Watts, equivalent to an annual consumption of around 4.5 MWh, in spite of the efforts which lead to a 3G BTS consuming less energy than a GSM BTS. It is clear that in view of these elements, beside the reduction in CO_2 emissions, operators are motivated by technologies for reducing energy consumption which would enable them to reduce their OPEX (*OPerating EXpenditure*). This objective is all the more crucial because cellular network operators are seeing an explosion in the volume of data traffic on their 3G/3G+ networks – and soon to be 4G/LTE networks – while their revenues are not growing in proportion with this, which is forcing them to reduce both the costs of their networks (CAPEX or CAPital EXpenditure) and the OPerating EXpenditure (OPEX).

Most of the techniques developed and implemented to reduce energy in mobile networks are based on parameters such as the data rate, QoS, availability, scalability, etc. Hence, these technologies attempt to overcome the main disadvantages to traditional – i.e. non-green – networks. Most of the techniques developed by designers of mobile communications systems were intended to maximize performance metrics such as data rate, QoS and reliability, and were not concerned with the energy consumption of the equipment of the network. Their design was performance-oriented rather than energy efficiency-oriented. Hence, gains in energy made with many green techniques could lead to a decrease in QoS. The compromise between energy consumption and the performances of the network is therefore a new problem to be considered in the design of such systems.

Figure 3.1 summarizes the sources of gains in spectral efficiency (and therefore energy consumption *in fine*) in radio communication systems [WEB 07].

The GSMA (GSM Association) took an interest in comparing mobile networks from the point of view of their energy efficiency. It defined four indicators that allow the comparison of networks: the network's consumption per connection, per cellular site, per unit traffic and per unit revenue.

The rest of this chapter is organized as follows: the first part is devoted to the processes and protocols implemented in green networks to optimize energy consumption. The second part relates to engineering aspects, and the third part looks at the aspects relating to hardware and architectures of the network elements to reduce their consumption.

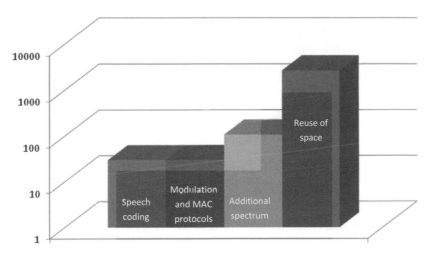

Figure 3.1. *Sources of gains in spectral efficiency in wireless communications systems between 1950 and 2000*

3.2. Processes and protocols for green networks

We can see in Figure 3.2 that out of all the components of the network, it is the radio access subsystem, and more specifically the base transceiver stations that consume the greatest amount of power in a cellular network. Indeed, the base transceiver stations consume between 50 and 90% of the energy of a cellular network. Thus, it is important to reduce the consumption of the radio access network (RAN) by any means possible. In fact, it is possible to reduce RAN emissions by 30%, and emissions in the core of the network by more than 50% by implementing the appropriate power-saving techniques. Let us focus here on the techniques employed to reduce the power in the RAN.

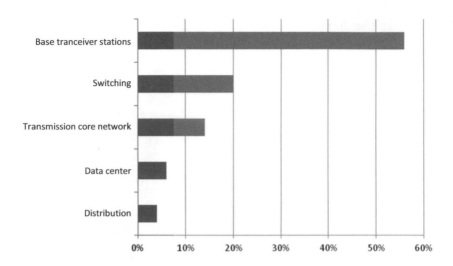

Figure 3.2. *Distribution of power consumption for a cellular network operator [CIO 08]*

3.2.1. *Technologies on the radio interface*

The technologies employed in the design of the radio interface transmission system includes techniques which we shall call "classic", such as power control, DTX, DRX and sleep mode, and more modern techniques such as MIMO, cognitive radio and interference cancellation.

Indeed, as pointed out in the introduction, problems relating to energy (power limitations of mobiles with a finite battery life, and an ever-smaller terminal size) have always been one of the major concerns for designers of systems, and in particular of the transmission channel. Hence, numerous techniques are elaborated and integrated into the systems' standards. Here we shall touch upon the two types of techniques put into practice: techniques that have been used since the time of second-generation systems and more recent techniques integrated in third- and fourth-generation systems.

3.2.1.1. *Power-saving techniques in 2G networks*

One of the first techniques to be put into practice in GSM for vocal communications is that of DTX (Discontinuous Transmission), which consists of not transmitting when the connected users are not speaking. On average, DTX makes it possible to save 60% of the power of emission from the terminals, because a voice communication occupies the transmission channel for around 40% of the time on average.

During periods of inactivity, the mobile also consumes power so as to "listen" to the network, particularly in mobility management and calls (typically to be informed of an incoming call). Hence, another technique introduced with GSM is DRX (Discontinuous Reception) which enables the mobile to monitor the broadcast channel and the paging channel of the current cell only at certain time intervals, predefined between the network and the mobile. Thus, it can stop receiving the rest of the time and save its battery. The same principle is implemented in WiMAX mobile networks (IEEE 802.16e) with the sleep mode which also allows the mobile to deactivate reception for periods of time fixed by the network. This enables power savings of between 25 and 36% [MAS 10].

3.2.1.2. *Power-saving techniques in 3G and 4G networks*

The MIMO (*Multiple Input, Multiple Output*) technique, defined and integrated into the 3GPP and WiMAX standards, for instance, enables the transmission and/or reception of different data flows transmitted or received by several antennas. MIMO can be combined with antenna lobe adaptation techniques, enabling the implementation of Space Division Multiple Access (SDMA), space-time coding and the HARQ retransmission technique. [BOU 07] shows that, thereby, energy efficiency can be improved by 30% in comparison with a non-adaptive MIMO system.

3.2.2. *Adaptation of network activity to traffic*

In second- and third-generation cellular systems, the operation of the network is independent of the state of the traffic. This means that all network elements are active even if the load only requires a reduced capacity, or even a null capacity at certain times and in certain areas. Thus, beginning with the observation that only 10% of infrastructures are actually used to carry traffic – a number which, when taken on a worldwide scale represents 80% of energy, equating to 100 million megawatt hours – optimization of the operation times of the equipment used in ICT is one possible avenue for reducing energy consumption. In [OH 10], it is stated that the percentage of time for which the traffic is below 10% of the maximum traffic during the day is around 30% on weekdays and 45% on other days. A base transceiver station with little or no activity consumes over 90% of its maximum energy (for instance, a UMTS BTS consumes 800-1,500 watts for an output power of 20-40 watts).

The consumption of a BTS increases exponentially with the binary rate:

$$Pt = a.(2R - 1)$$

where Pt is the power transmitted, R the binary rate and a the factor representing the quality of the radio link.

The advantage of reducing the consumption of BTSs is also fuelled by financial imperatives, as figures show that traffic is increasing by 400% per year [CHI 08], in contrast to an increase in revenue of only 23%. Hence, savings must be made both at the level of the usage loads and at the level of investment.

Thus, equipment manufacturers have introduced mechanisms which enable the power consumption of transmitters to be dynamically adapted to the traffic. The power of each TRX (transceiver or transmitter-receiver) can be adjusted in a time-slot of a millisecond in the case of GSM, for example, thereby facilitating reductions in power consumption of between 25 and 30% over a 24-hour period. This principle of traffic adaptation, in combination with the use of low-power processors, means the energy efficiency can reach up to 70% at the level of the core nodes of the network. For the WCDMA technique which is used in UMTS, the equivalent of this technique is to deactivate carriers when traffic decreases and that a single carrier is sufficient to handle the available traffic.

By way of an example, installing a million GSM base transceiver stations with this function would reduce CO_2 emissions by around a million tons a year (the equivalent of 330,000 cars driving 16,000 km per year).

The use of this technique based on fluctuations in traffic to minimize energy consumption by turning off the base transceiver stations or TRXs [SAK 10] is called dynamic planning.

3.2.3. *Traffic aggregation based on the delay*

When the QoS allows it – which is the case, notably, for applications with elastic QoS such as interactive services like the Web or background services like e-mail – it is possible to pack the data to be transmitted together, in order to reduce the overhead transmission costs of each element of traffic, and also to reduce the consumption of transmitters/receivers by reducing the number of transmissions/receptions. The technique put in place, for a transmitter, therefore consists of waiting until it has received a certain volume of data to transmit from the higher layers before actually beginning to transmit. In the WiMAX system, for instance, an aggregation technique is defined which enables us to accumulate small packets at layer 2. A tradeoff between the energy savings made and the degradation of the QoS is defined for managing this technique.

3.2.4. *Store, carry and forward relaying*

In [KOL 10a] and [KOL 10b], the data transfer technique called Store, Carry and Forward Relaying (SCF) is proposed for non-real-time services. Based on the fact that in cellular networks, most of the traffic is carried by a small number of sites, whereas other cells remain under-used, the authors suggest the SCF technique, which allows base transceiver stations with low traffic load to be turned off during periods of low traffic and to relay the traffic in their coverage area using other base transceiver stations. The SCF technique relies on mobile nodes to carry the traffic from cells whose base transceiver stations have been deactivated to active cells. As is the case with the previous technique, it is clear that a tradeoff between the possible gains in terms of reduction of the power consumed and the delay in transmission must be taken into consideration. The authors show that with an acceptable delay of between 10 and 30 seconds, and for between 10 and 20 vehicles, the gains vary from 20% to nearly 90% in terms of energy saved.

3.2.5. *Combination of MS and BTS*

Since the MS-BTS radio link is the segment which consumes the most power in a transmission link between two points, optimizing the power involved enables gains to be made over the whole system. Hence, it is essential to focus on designing techniques for determining the optimal MS-BTS coupling for a particular set of radio propagation conditions between the two points. The main objective of the handover algorithms – in the target cell selection criteria – is to identify the optimal target for the intercellular transfer (or handover). These algorithms mainly use the power of the link as the primary criterion, as in the case of the "best cell" criterion for the handover or the FBSS (Fast Base Station Switching) algorithm defined in WiMAX.

3.2.6. Handover for optimization of the energy used

A particular case of the above technique is the handover algorithm, which is particularly important in areas covered by many RAT (Radio Access Technology) installations. Taking account of the constraints relating to the bandwidth of the service established, and of the energy efficiency of the link, enables us to save energy. The two types of handover by which these objectives can be attained are vertical handovers and horizontal handovers (see Figure 3.3). Vertical handovers enable transfers between different types of technology, with the aim of ensuring the best possible connectivity to mobile applications, transparently and without call drop. In the case of horizontal handover, transfers take place between cells of the same technology or the same layer. The main factors taken into account by the handover algorithms are the QoS, the cost of service, etc. [HAS 05]. Secondly, the factors based on energy consumption in handover that best target cell identification are recounted in [CHO 07; YAN 09; SEO 09; PET 09].

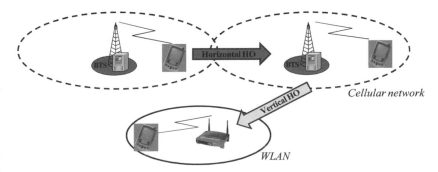

Figure 3.3. *Illustration of the mechanisms for vertical and horizontal handover between cellular networks and WLAN*

3.2.7. *Cooperation between base transceiver stations*

In hot spot areas (i.e. with dense traffic), each point can be covered by several base transceiver stations at once. In a situation where there is heavy traffic, all the cells are needed for the traffic flow. Conversely, in a low traffic period, certain base transceiver stations are surplus to requirements and it is possible to deactivate them (see Figure 3.4), the result of which is a reduction in their energy consumption. Coordination between base transceiver stations enables us to determine which base transceiver station should remain active and which should be deactivated, so as to cover the active terminals while minimizing the overall power consumption in the area.

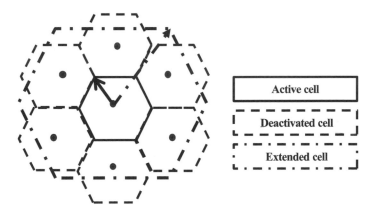

Figure 3.4. *Dynamic deactivation of cells in a cellular network*

3.2.8. *Increasing the capacity of the RAN and network core nodes*

The efforts made in terms of reducing the number of sites deployed led to the design of BSCs (Base Station Controllers) with high capacity (for example, capable of handling over 4,000 TRXs and 25,000 Erlangs of traffic) in a single location, which leads to an 80% saving in terms of energy consumption. 3G networks, with RNCs (*Radio Network*

Controllers) are also affected by this evolution. In fourth-generation networks, based on flat architecture, by removing the base station controllers the gains obtained at the level of the RAN are increased, enabling us to reduce the OPEX and CAPEX. Finally, note that in 4G networks as well, reduction to a single type of network subsystem such as a packet, enables this trend to be extended, with core networks essentially relying on routers which are more compact and less energy hungry than the circuit switches used in 2G and 3G cellular networks.

3.3. Architecture and engineering of green networks

The second axis along which the efforts to design green networks are oriented is that of the architecture of the networks. Indeed, if we adopt an optimized and appropriate architecture, energy savings can be made. Below, we present the main architectures identified in this context.

3.3.1. *Relaying and multi-hopping*

The multi-hop mobile networks considered in the 3GPP standards enable a link between a mobile and base transceiver station to be obtained by way of several terminals (a single hop in the case of LTE). One notable advantage to this, besides the improvement of the QoS and coverage of a site, is a reduction in power because the transmission links are shorter for each hop. Relay nodes (RNs) are thus defined, which enable coverage to be offered to smaller areas than macrocells, and consequently require far less transmission power in comparison to eNode Bs. RNs manufactured for smaller ranges therefore constitute a very interesting solution to improve the energy efficiency of a mobile network. Different types of RN are suggested by the 3GPP (3G Partnership Project). One such type is in-band relays, which use the same frequency band as the links between RNs and base stations and those between RNs and mobiles. Below, we present the two types of relays defined by the 3GPP:

– Type 1 relays: these relays operate at layer 3, meaning that the protocols up to layer 3 for the user packet data are embedded in the RN [3GP 09]. A level-3 relay thus includes all the functions of a base transceiver station and can receive and transmit IP packets (PDCP SDUs). Hence, these relays are visible for mobile terminals because they integrate layer 3;

– Type 2 relays: these relays may integrate either the protocols up to layer 2 or up to layer 3, depending on the solution implemented. In 3GPP terminology, type 2 relays are transparent for the mobile terminal.

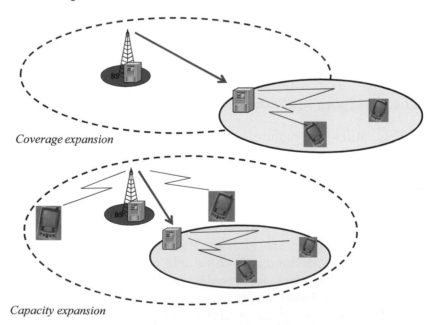

Figure 3.5. *Approaches to the use of relay nodes*

Relays also ensure two major functions in cellular networks: the extension of coverage outside the area covered by a base transceiver station, and the extension of the capacity of a base transceiver station (see Figure 3.5).

Thus, the installation of relay nodes in the network involving multi-hop transmission techniques enables us to reduce the energy needed for a base transceiver station to mobile communication. Nevertheless, this technique exhibits the disadvantage – as do the techniques introduced above – of the possible degradation in the QoS, particularly in terms of the delay parameter or latency, due to the addition of an intermediary node into the transmission channel.

3.3.2. *Self-organizing networks (SONs)*

Self-organizing networks, or SONs, are able to react automatically, by various mechanisms for adapting their parameters and configurations, to the constraints and conditions of traffic, QoS, breakdown, etc. They stem from the increase in the number of elements in the network and the resultant impossibility for a human being to configure and optimize all these components. Manual techniques have been progressively replaced by SONs. These techniques are focused on:

– how to automate the configuration and establishment of the network (*self configuration*);

– how to optimize the parameters of the network so as to obtain better performances (*self optimization*);

– how to recover the performance of service in the case of breakdown or dysfunction of the network (*self healing*).

Thus, an SON is a network with the capacity to automate the management processes. This function enables us to minimize the operational lifecycle of a network by eliminating the manual configuration of the equipment during installation, directly and dynamically during the function of the network. The ultimate goal is to reduce the costs and prices of wireless data services [NOK 09].

3.3.3. *Planning*

Coverage network planning may be one way of economizing on power and thereby respond to the objective of green networking.

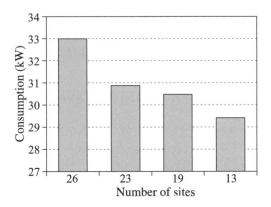

Figure 3.6. *Reduction of consumption by an optimized network design*

In second-generation networks and in the early days of third-generation networks, the high costs induced by radio sites and the associated equipment were largely compensated by the savings that it was possible to make in other areas of expenditure, such as transmission equipment, civil engineering, network rollout, operation and maintenance, location of sites, transport, spares, support and training. The pressures to make savings at the level of the network, either in CAPEX or OPEX, are currently leading designers to propose network structures which enable us to save on radio sites but also on their energy consumption. This aim can be served by optimizing the engineering of the radio sites by choosing appropriate values for the parameters such as position (where possible), height, direction (azimuth), diagram (type) and inclination (tilt) of the antennas, transmission power, relative mutual coverage of sites. Thereby, substantial savings of several tens of percent can be achieved. The location of the sites in relation to the traffic is a problem which has arisen with CDMA technology for reasons specifically linked to the access technique. Indeed, because the base transceiver station shares its power amongst all the mobiles within range of the cell, the problem of minimizing the power of each link becomes more restrictive in CDMA than in TDMA, for instance, where all the power of the base transceiver station is available for the mobile to which that time slot is allocated. The problem of minimizing power in the context of green networking is an additional reason to plan the radio network so that the sites – or at least the antennas of the base transceiver stations – are as close as possible to high-traffic areas. The power needed to serve a terminal will thus be reduced, because of a best cell link which is more favorable thanks to more favorable mobile-to-BTS propagation conditions. This type of planning is known as traffic-aware network planning, and also offers substantial gains at the level of power consumption for each link considered separately, and gains at the global level because the overall level of interference has been reduced.

3.3.4. *Microcells and multi-RAT networks*

In the same type of problems as optimization of the radio engineering, reducing the radius of a 1,000 m cell to 250 m increases the efficiency of the system from 0.11 Mbits/Joule to 1.92 Mbits/Joule, which equates to a 17.5-fold gain as evaluated by Badic *et al.* [BAD 09]. Indeed, the radio link loses 12 dB when the radius doubles, with an attenuation coefficient of 4.

The use of microcells in macrocell networks has the advantage of not only increasing the capacity of the network but also facilitating a reduction in energy consumption. In fact, the greater the required capacity, the more cells that are needed, of reduced size, yielding significant gains, as Figure 3.6 shows. The higher the number of microcells, the more the power consumption in the area can be reduced. In [EAR 10a], the volumes of traffic considered for LTE range between 1 Mb/s/km^2 (rural areas with low traffic) and 120 Mb/s/km^2 (dense urban environment with high usage). In [EAR 10b] it is shown that, for a capacity of 70 Mb/s/km^2, coverage by macrocells is the optimal scenario. For traffic densities of up to 100 Mb/s/km^2, one microcell per macro site is the optimal deployment scenario, and in this case facilitates gains of up to 3% in relation to a scenario where only macrocells are used. At traffic densities greater than 150 Mb/s/km^2, deployments including at least five microcells per macro site become advantageous, with potential gains of up to 10%. It should be noted that these results apply in the case of traffic uniformly distributed across the surface area being covered and that the microcells are placed on the boundary of the macrocell (i.e. in areas where coverage is not as good). Another interesting argument in favor of microcells is that where the cells are placed inside the macrocell. Note, finally, that it only makes sense to use microcells for heavy traffic and high data rates of transmission, positioning them as an alternative to macrocell networks from an energy standpoint for traffic densities beyond 250 Mb/s/km^2 [MAR 03; HOP 99]. The deployment of femtocells in indoor environments, and traffic-carrying by xDSL wireless routers or by micro-BTSs are also involved in reducing the power used on the mobile-to-network link, using network elements connected to the base transceiver stations, to the controllers or directly to the switches or routers by wires, which consume less energy than wireless links do.

In addition to the type of cells used in the network, the frequency bands used can also help save on power. Thus, an access network operating with numerous frequency bands – for instance, in the 900 MHz band and in the 2,100 MHz band – facilitates higher data rates and therefore greater capacities. Indeed, the mobile-to-BTS link can be optimized using the additional parameter of the frequency band (low-band in hard propagation conditions and high-band for good or excellent conditions which allow for higher data rates). The results presented in [EAR 10b] show gains of 15% attainable in terms of the power used, as well as in terms of capacities.

3.3.5. A step towards all-IP and flat architecture

Fourth-generation networks are characterized by a flat architecture, consisting of a single layer of base transceiver stations, mutually interconnected by large broadband links (generally fiber optic). Eliminating the layer of controllers (BSC, Base Station Controller and RNC, Radio Network Controller) present in 2G and 3G access networks results in a reduction of the costs (linked to deployment, ground rent on the sites, etc.) and energy consumption (particularly with more energy-efficient and higher-capacity equipment). In addition, the elimination of the circuit-switched (CS) core network in 4G networks enables us to pursue this trend and reduce the number of nodes, and consequently the energy consumed by the network, in particular with more power-efficient routers.

3.3.6. Reducing the number of sites by using smart antennas

Smart antennas were introduced into third- and fourth-generation networks. These antennas facilitate the use of techniques such as MIMO (Multiple Input Multiple Output) and beamforming, in combination or otherwise. Thanks to these smart antennas, extended coverage techniques with sites enabling larger areas to be served, are available in new networks. The technique of beamforming initially introduced into WiMAX networks and taken into account in the 3GPP standards (for WCDMA and LTE systems) focuses the energy in one particular direction (the direction of the target mobile), which reduces the radiation (and therefore the energy and the interference generated) required for the network-to-mobile link. This particular technique allows gains of around 40% in terms of the number of sites.

3.3.7. *Cooperation between BTSs*

One of the main objectives of cooperation between BTSs is to ensure that all the energy consumed by the base transceiver stations is used efficiently for data transport. Uncontrolled interference causes wastage of energy. Thus, several methods for cooperation between base transceiver stations have been implemented, and are generally denoted by the umbrella term coordinated multi-point (CoMP) communication. These methods set up communication between neighboring base transceiver stations by backhaul links. Depending on the capacity of these links, cooperation may take place at the terminal level or at the network level by coordination of the resources allocated to the terminals and minimizing the inter-BTS interference.

The CoMP technique offers gains in terms of energy consumption [AKT 06] as regards the link between the base transceiver stations and the terminals, particularly when the quality of the link is mediocre or when the level of interference is high (that is, typically for users located on the boundary of a cell's range). In addition, the results show that cooperation between at least three base transceiver stations facilitates an improvement in energy consumption per bit. While communications carried out on the boundary of a cell (i.e. within 5 to 10% outside the cell's signal radius) correspond to 10-20% of the traffic load, distributed uniformly, the reduction in energy consumption would be around 5-10% [EAR 10b]. In [STO 08], frequency reuse techniques based on fractional patterns show similar gains. The gains may reach up to 20% when the load is low and 12% when it is high, based on the following hypotheses concerning traffic [EAR 10b]: low traffic if for nine hours a day, the traffic is lower than 50% of the maximum daily traffic, and high traffic if for 15 hours a day, the traffic is greater than 50% of the daily maximum. The overall gain of the system is around 15% in these conditions.

Note, finally, that in [MAN 10], the authors show that the level of redundancy of a network enables the power consumption to be reduced. They suggest an approach based on cooperation between the base transceiver stations so as to minimize the number of active BTSs while satisfying the minimum level of QoS and the coverage required.

3.4. Components and structures for green networks

Most of the techniques used to minimize energy consumption in the context of green networks attempt to optimize the transmission power of the base transceiver stations using engineering or transmission chain technologies. Here we introduce the measures which are implemented at the equipment level – i.e. the level of the components and structures of the networks. As specified in the introduction, the elements of the network whose consumption accounts for the majority of the system's power usage are those of the radio access network. Remember that the base transceiver stations consume a high proportion of the energy in cellular networks (between 60 and 80%). This illustrates the importance of reducing the power consumed by BTSs. Also, besides the environmental effect sought, reducing the consumption of base transceiver stations also has indirect consequences for all sorts of savings, such as the prolongation of battery life (mobiles and BTSs) and therefore all related costs (maintenance, replacement, etc.).

The energy used in base transceiver stations can be reduced in various ways:

– by developing innovative components such as energy-efficient amplifiers, fanless air-conditioners, or even components which can operate at high temperatures without cooling;

– management of resources, such as power control;

– smart network topologies, from deployment to operation (e.g. by the use of relays, dynamic site switching, etc.).

Some of these techniques have been introduced in the previous sections. Below we present the techniques relating to energy-efficient components and alternative energy-producing technologies.

3.4.1. *Power-efficient amplifiers*

Base transceiver stations contain a radio-frequency (RF) component, which consumes a large proportion of the energy. The energy consumption of the RF part represents up to 40% of the total consumption of a BTS. The proliferation of the radio standards available on a radio site with the constraint of minimizing the power of base transceiver stations has led to the introduction of multi-mode base transceiver stations which can deal with several standards (GSM, W-CDMA, LTE or WiMAX). For this purpose, these BTSs include multi-carrier power amplifiers which offer power savings of up to 60%. Thus, by combining several technologies at the level of the amplifier, their efficiency may reach up to 40%.

Another technique is introduced by High Accuracy Tracking Power Modulators which, by replacing the DC/DC converter, allows us to dynamically alter the voltage produced by the power amplifier depending on the envelope of the signal, thereby offering power gains of around 50%.

3.4.2. *Elimination of feeders, use of fiber optics*

Typically, base transceiver stations integrate the RF stage within their architecture. Because the antennas are situated several or even dozens of meters away from the base transceiver station, the attenuation of the signal as it is transmitted along the supply cable (known as the feeder) linking the RF stage to the antennas may be all the more significant, and consequently necessitates an increase in the transmission power at the level of the BTS. Hence, nearly all recently-manufactured BTSs use amplifiers placed at the level of the antennas, thereby shortening the distance between the transmission and reception of the signal from the antenna, and minimizing the need for amplification before or after transmission of the signal through the feeder cable. The gain in power consumption thus attained can be up to 25%.

In addition, the use of amplifiers located near the antennas requires them to function at very high temperatures (up to 45°C) without cooling. In fact, designing components which can operate at high temperatures enables us to reduce, or even eliminate, the need for cooling, and consequently the energy consumption associated with it at the level of the radio sites.

3.4.3. *Solar and wind power*

Using green energies such as solar or wind power also reduces CO_2 emissions significantly. In certain countries where electrical energy is not available throughout the territory, the use of fuel-burning generators indirectly caused CO_2 emissions by the supply vehicles (once a month or once a quarter) which constantly circulate between radio sites equipped with generators and tanks to supply them with fuel. The use of solar or wind power as an energy source by a thousand sites saves 4,400 tons of carbon and therefore 14,500 tons of CO_2. In more concrete terms, one square meter of solar paneling produces 400 kilowatt hours of energy, which is around 10% of the average requirements of a 3G macrocell, and at least 5% (which is the case for dense networks such as London, for instance). The combination of solar and wind energy sources, on a site where both are feasible, would supply the needs of a micro- or picocell. In areas of low traffic, solar panels are already in use, and provide a constant supply to the base transceiver stations, particularly in sub-Saharan zones, which receive a great deal of sunshine all year round.

3.4.4. *Twin TRX*

The technology called twin TRX is championed by certain manufacturers. This involves physical TRX modules (transceivers that represent the radio stage in the BTS) which are each able to handle two virtual TRX modules. This enables the power consumption of each TRX to be reduced by a little more than 30%. An additional power saving of up to 60% can be obtained by using Multi-carrier Power Amplifier hardware.

3.4.5. *Cooling*

Energy savings can also be found in the environment of a radio site. We mentioned the case of power supply with the use of alternative energy sources. Cooling is another example of a very energy-hungry element in the environment of a base transceiver station. In data centers, for instance, 50% of electricity is consumed by the cooling components, the other

50% being used by the computing equipment [FAN 07; BAR 07]. Improving the efficiency of the cooling machinery and using alternative means of cooling make a tremendous contribution in terms of reducing the energy consumed.

3.5. Conclusion

In this chapter, we have presented the various approaches used to try to reduce energy consumption in cellular networks to achieve green networks. We have also seen that this objective is sometimes attained to the detriment of the QoS, and that consequently there has to be a certain compromise preserved between the power saving made and the degradation of the QoS offered by the services in question.

3.6. Bibliography

[3GP 09] 3GPP TR 36.814 v1.5.1 Further Advancements for E-UTRA, Physical Layer Aspects, December 2009.

[AKT 06] AKTAS D., BACHA M., EVANS J., HANLY S., "Scaling results on the sum capacity of cellular networks with MIMO links", *IEEE Trans. Inform. Theory*, vol. 52, no. 7, p. 3264-3274, July 2006.

[BAD 09] BADIC B., O'FARREL T., LOSKOT P., HE J., "Energy efficiency radio access architectures for green radio: large versus small cell size deployment", *Proceedings of the IEEE 70th Vehicular Technology Conference (VTC Fall)*, Anchorage, United States, September 2009.

[BAR 07] BARROSO L.A., HÖLZLE U., "The Case for Energy-Proportional Computing", *Proceeding Of IEEE Computer*, p. 33-37, 2007.

[BOU 07] BOUGARD B., LENOIR G., DEJONGHE A., VAN PERRE L., CATTHOR F., DEHAENE W., "Smart MIMO: an energy-aware adaptive MIMOOFDM radio link control for next generation wireless local area networks", *EURASIP J. Wireless Commun. Networking*, vol. 2007, no. 3, p. 1-15, June 2007.

[CHI 08] CHIA S., "As the Internet takes to the air, do mobile revenues go sky high?", *IEEE Wireless Communications and Networking Conference*, Las Vegas, April 2008.

[CHO 07] CHOI Y., CHOI S., "Service Charge and Energy-Aware Vertical Handoff in Integrated IEEE 802.16e/802.11 Networks", *Proceeding of IEEE INFOCOM*, p. 589-597, 2007.

[CIO 08] CIOFFI J.M., ZOU H., CHOWDHERY A., LEE W., JAGANNATHAN S., "Greener copper with dynamic spectrum management", *Proceeding of IEEE GLOBECOMM*, p. 1-5, 2008.

[CIS 10] CISCO, Cisco visual networking index: global mobile data traffic forecast update, 2010-2015, 2010.

[EAR 10a] EARTH PROJECT DELIVERABLE D2.3, Energy Efficiency Analysis of the Reference Systems, Areas of Improvements and Target Breakdown, 2010.

[EAR 10b] EARTH PROJECT DELIVERABLE D3.1, Most Promising Tracks of Green Network Technologies, INFSO-ICT-247733 EARTH, December 2010.

[FET 08] FETTWEIS G., ZIMMERMANN E., "ICT energy consumption trends and challenges", *Proceedings of IEEE WPMC*, Lapland, Finland, September 2008.

[HAS 05] HASSWA A., NASSER N., HASSANEIN H., "Generic vertical handoff decision function for heterogeneous wireless networks", *Proceeding of IFIP Conference on Wireless and Optical Communications*, p. 239-243, 2005.

[HOP 99] HOPPE G.W.R., LANDSTORFER F.M., "Measurement of building penetration loss and propagation models for radio transmission into buildings", *IEEE Veh. Technol. Conf.*, p. 2298-2302, September 1999.

[KAR 03] KARL H., An overview of energy efficient techniques for mobile communications systems, Technische Universität Berlin, Technical Report, 2003.

[KEL 07] KELLY T., ICTs and climate change, ITU-T Technology, Technical Report, 2007.

[KOL 10a] KOLIOS P., FRIDERIKOSY V., PAPADAKI K., "MVCE green radio project: inter-cell interference reduction via store carry and forward relaying", *Green Wireless Communications and Networks Workshop (GreeNet)*, VTC Fall, 2010.

[KOL 10b] KOLIOS P., FRIDERIKOS V., "Load balancing via store-carry and forward relaying in cellular networks", *IEEE GLOBECOM*, 2010.

[MAN 10] MANCUSO V., ALOUF S., "Reducing costs and pollution in cellular networks", in *IEEE Communications Magazine*, special issue on Green Communications, June 2010.

[MAR 03] MARTIJN E.F.T., HERBEN M.H.A.J., Characterization of radio wave propagation into buildings at 1800 MHz, Thesis, 2:122-125, 2003.

[MAS 10] MASONTA M.T., MZYECE M., NTLATLAPA N., "Towards energy efficient mobile communications", *Proceedings of the 2010 Annual Research Conference of the South African Institute of Computer Scientists and Information Technologists, SAICSIT*, 2010.

[NOK 09] NOKIA SIEMENS NETWORKS, Introducing the Nokia Siemens Networks SON suite – an efficient, futureproof platform for SON, November 2009.

[OH 10] OH E., KRISHNAMACHARI B., LIU X., NIU Z., "Towards dynamic energy-efficient operation of cellular network infrastructure", *IEEE Communications Magazine*, November 2010.

[PET 09] PETANDER H., "Energy-aware network selection using traffic estimation", *Proceeding of ACM MICNET*, p. 55-60, 2009.

[SAK 10] SAKER L., ELAYOUBI S.E., CHAHED T., "Minimizing energy consumption via sleep mode in green base station", *Proceeding IEEE WCNC*, p. 1-6, April 2010.

[SEO 09] SEO S., SONG J., "Energy-efficient vertical handover mechanism", *Proceeding of IEICE Transactions on Communications*, vol. E92-B, no. 9, p. 2964-2966, 2009.

[STO 08] STOLYAR A.L., VISWANATHAN H., "Self-organizing dynamic fractional frequency reuse in OFDMA systems", *INFOCOM 2008, The 27th Conference on Computer Communications IEEE*, p. 691-699, 13-18 April 2008.

[WE 07a] WEBB W., *Wireless Communications: The Future*, Wiley, New York, 2007.

[WE 07b] WEBER W.D., FAN X., BARROSO L.A., "Power provisioning for a warehouse-sized computer", *Proceeding of ACM International Symposium on Computer Architecture*, p. 13-23, 2007.

[WIL 08] WILLIAMS F., Green wireless communications, eMobility, Technical Report, 2008.

[YAN 09] YANG W.H., WANG Y.C., TSENG Y.C., LIN B.S.P., "An energy-efficient handover scheme with geographic mobility awareness in WiMAX-WiFi integrated networks", *Proceeding of IEEE WCNC*, p. 2720-2725, 2009.

Chapter 4

Green Telecommunications Networks

4.1. Introduction

The world of networks, and information technology (IT) in general, requires a considerable amount of energy, which is often underestimated. The consumption at the beginning of 2012 was revealed to be approximately 5% of the worldwide carbon footprint. This value is difficult to calculate, and estimations place it at between 3 and 7%. This consumption is due to numerous factors, which we shall study in this chapter. To begin with, the ratio of the consumption of IT to that of telecommunications is roughly 2:1.

Before taking a more in-depth look at the world of telecommunications, let us highlight that in IT, five-sixths of the energy expenditure goes on personal computers, printers and all the other associated equipment. Several billion sockets are needed to supply these machines with power, which are often on 24 hours a day and consume between 20 and 100 Watts. The final sixth relates to the data centers,

---

Chapter written by Guy PUJOLLE.

i.e. the computer servers located in data warehouses which constitute the memory of the Internet, and increasingly its computing power and its applications. The largest data centers sometimes contain a million servers, and their electrical consumption can be up to 100 MW.

Data centers form the neural points of the Cloud, and more precisely, the Clouds of Internet service providers (ISPs), storage, computation, etc. The large Cloud providers sometimes have several data centers for reasons of availability, reliability and redundancy. In section 4.2, we shall examine the elements which might help reduce the electricity consumption of these monsters.

The second major domain which we are going to look at in greater detail is the consumption of the networks; the Clouds could be seen as the servers associated with those networks. If we do not look at it in excessively great detail, we see that the consumption stems largely from the antennas of telecommunications operators. A third-generation (3G) antenna can consume between 1,000 and 2,000 Watts, with an average consumption of 1,300 Watts. If we multiply this figure by the number of antennas, the result soon becomes astronomical. There are nearly 2 million antennas in the world. Large countries such as India have up to 13 independent operators.

A second major source of expenditure relates to Home Gateways, which are usually called Internet boxes, broadband routers or ADSL routers. These are so numerous they are normally counted in hundreds of millions; there are 500 million worldwide. While the energy consumption for each box is not enormous (between 10 and 30 Watts), the total when taken for all the boxes in existence is considerable to say the least.

Following this brief introduction about the energy consumed by the world of ICT, we shall now go into a little more detail about the reasons for this consumption, but above all about the possible solutions to reduce it. We shall begin our examination with data centers and Clouds, whose significance becomes critical in the use of telecommunications networks; then we shall give a detailed analysis of green technologies for the world of networks itself.

4.2. Data centers

Data centers are vast arrays of servers which store the data of individuals and companies that use Clouds. The largest are those owned by Google, Amazon, Microsoft and Facebook. It is estimated that the number of servers increases by 20,000 every day. In the knowledge that the average consumption of a server is around 30 W, this represents 600 kW extra each day. In terms of the cost, in 2012, the cost of electricity supply equates to approximately half the total cost of running the data centers. Figure 4.1 traces the evolution of these costs in billions of dollars.

Between 2008 and 2009, the cost of the energy provision became equal to the cost of the hardware.

The distribution of electrical consumption in data centers is described in Figure 4.2. Around 30% of this consumption is attributable to the computing equipment as such – that is to say that only a third of the electricity is used to run the servers. Cooling occupies a second third, and finally, in order of consumption, we find UPSs (Uninterruptible Power Supply) – systems which guarantee an absolute constant electrical current – and CRACs (Computer Room Air Conditioner), which are "smart" air conditioners capable of adapting to the need for cooling.

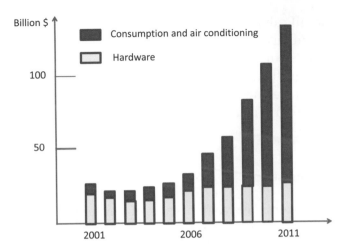

Figure 4.1. *The cost of data centers*

There are numerous potential avenues for energy saving in the domain of data centers. We shall briefly describe a few of these. The first area in which a saving could be made is cooling. By situating the data centers in cold countries, thereby using "free cooling" – i.e. using the cold air from outside – the costs of cooling would obviously be massively decreased. Another solution is to use the hot air produced by the data centers to heat dwellings or commercial premises in the vicinity.

As regards the computer processors, they are often not well optimized in relation to the work to be carried out. Virtualization is one significant economizing factor. Virtualization consists of creating software suites to perform tasks which were previously done by hardware: one simply describes the task carried out by the hardware in the form of a computer program, and executes that code on a sufficiently powerful machine to yield the same performances as the physical machine. *A priori*, we see a fairly large loss, because the computation is less optimized and there is a need for a hypervisor to manage the different virtual machines.

However, the advantages to virtualization are many: to begin with, the virtual machines can be moved from one physical machine to another.

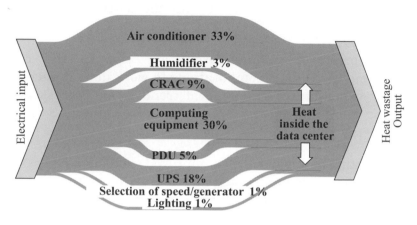

Figure 4.2. *Electrical consumption of data centers*

Thus, it is possible to host a large number of virtual machines on the same physical machine, until that physical machine is being used to the best of its capacity. Given that, worldwide, the average usage of servers is around 10%, virtualization should (at least, in theory) enable us to switch off 9 out of 10 machines, on average. In reality, the gain is considerably less, because the server cannot reach its upper limit – the hypervisor must be taken into account, and finally, availability and reliability must be ensured by having redundant resources to take over in case of failure. If virtualization is used properly, we can speak of a five-fold gain. In fact, the processors are put to far better use.

Another solution which is the focal point of many research projects and developments relates to the speed of the processor, which ought to be able to adapt its processing power to the task needing to be done. The less work there is on the server, the more the speed of the processor should decrease. Today, most machines have processors that are capable of working at different speeds, but in general, they do not adapt automatically to the workload. It is the user who decides whether or not to put the machine in energy-saving mode.

Microprocessor manufacturers are also in the race, improving the ratio of computations to work done in terms of energy expenditure. The general trend is for processors to become more and more powerful while consuming less and less energy.

Program channels are also improved, to enable processors to maintain a constant speed and therefore use the physical resources far more efficiently.

Overall, we can conclude that there is now a greatly raised awareness of the issue of energy expenditure, and numerous measures have been taken to attempt to minimize it, or at least to use the energy in a much more efficient way.

4.3. Wireless telecommunications networks

Telecommunications networks also devour energy. This is due to both the TCP/IP protocol itself and to the hardware elements used in this sector. The TCP protocol sets a very bad example as regards electrical consumption. For a very long time, there was no problem, because the machines were connected to high-power electrical sockets. Since the advent of mobile phones, smartphones and ever-smaller receivers, awareness of the electrical consumption of the TCP protocol has increased. This consumption comes directly from the protocol itself. There are a number of ways in which this energy expenditure can be determined. For instance, transmitting packets one after another on an antenna may be a very energy-hungry process, whereas by grouping the packets together so as to only send one long packet, the energy consumption of the transmitter can be at least halved. Another example relates to the number of timers to be triggered in order to manage "slow-start" in Internet network control techniques. There are an abundance of examples to show the energy inefficiency of the TCP/IP domain. A number of working groups of the IETF have been devoted to this issue since the turn of the millennium, but no great progress has been made.

To begin with, let us turn our attention to the weak points of consumption in telecoms networks. As was pointed out above, one of the main vectors of consumption relates to the relay antennas of mobile operators: base transceiver stations (BSTs) or "node B" posts. Figure 4.3 offers an estimation of the consumption of these antennas.

We can see in Figure 4.3 that over half the consumption is due to the power amplifiers. The next-biggest culprit is cooling, at around 20%, followed by signal processing with around 10%, and another ~10% for electricity supply. A great deal of progress can be made in this domain.

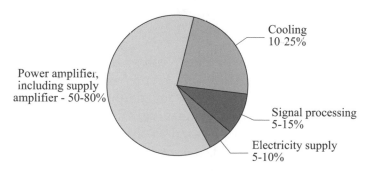

Figure 4.3. *Breakdown of the consumption of relay antennas*

Speaking only of the largest proportion of consumption, which is that of the signal amplifier, it should be noted that the power of the signal is proportional to the square of the distance to be covered. Thus, the further an antenna has to transmit, the higher its power must be. It is fairly easy to see that if two machines communicate with one another over a long distance, it will be very economical to install an intermediary relay. If L is the distance (length) to be covered, we have $L^2 > (L/2)^2 + (L/2)^2 = L^2/2$. This calculation is an approximation, of course, because an intermediary antenna must be installed and powered, but the overall gain made will be significant. It is for that reason that the wireless communications of the future should have smaller and smaller ranges. However, the installing of intermediary relays poses numerous problems – particularly local residents' refusal to have them near their homes. This is one of the arguments in favor of femtocells, which are antennas that are built into Home Gateways. They would be extremely numerous and would only serve small areas.

Another major source of consumption is the data rate of telecoms antennas, which is simply increasing with the exponential growth in demand around the world, due mainly to smartphones, which are always connected and work throughout the day. In Figure 4.4, we have illustrated the increase in data rates in recent years and the projected increase for years to come.

This figure shows data rates doubling every year. Forecasters predict that this trend will continue until the 2020s, which is ten years of doubling, representing a thousand-fold increase in data rates. Given that electrical consumption is closely related to the data rate, this implies at least a hundred-fold increase between 2010 and 2020. This increase will be achieved by way of various technological innovations, such as cognitive radio, which enables better use to be made of the spectrum, harvesting idle frequencies to broadcast with the obligation to stop doing so if the primary source on that frequency begins broadcasting. For instance, the future WiFi IEEE 802.11af will use unoccupied television frequency bands (known as white spaces) to broadcast at high data rates, using a far wider portion of the spectrum. Given that in most rooms the walls prevent good reception of television channels, this cognitive radio technology using television frequencies will be particularly effective within rooms. Another reason for this increase in capacity is the use of directional antennas, which are able to transmit only in the direction of the receiving antenna instead of omnidirectional broadcast, i.e. transmitting in all directions at once.

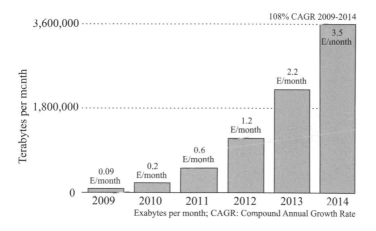

Figure 4.4. *Increase in data rates in the world of wireless*

This increase in data rates is due to video applications, and to a lesser extent, to distributed gaming. These applications are indicated in Figure 4.5.

We can see in this diagram that video takes the lion's share, occupying two thirds of the rates, followed by Web access and data transmissions, particularly to Cloud applications. P2P plays a far smaller role than one might imagine, followed by mobile gaming and finally telephone calls, which account for only 4% of the bandwidth, which corresponds fairly closely to recent years, even if the financial share occupied by telephone calls represents far more than these 4%.

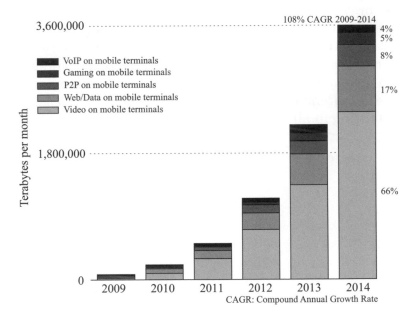

Figure 4.5. *Data rates by application in the world of wireless*

Energy savings are made by better use of the physical machines. The examples of potential reductions in energy are legion. One of the most significant is that of virtualization, for data centers, for instance. Virtualization performs the task of a physical machine in software form, and such virtual machines can be grouped together on a single, shared physical machine. For example, it would be perfectly possible to assemble all the operators on a single physical antenna by virtualizing them in terms of data processing capacity. Above a hypervisor, each operator could have its own virtual machine, as shown in Figure 4.6. The operators have complete ownership of their virtual machines and can therefore program them as they see fit. The advantage is that the various services then share common hardware which is far more powerful but which does not really consume much more energy. In addition, a great many pieces of hardware could be shared, such as VLRs (*Visitor Location Register*) or HLRs (*Home Location Register*). This solution would reduce the number of antennas and drastically decrease electrical consumption.

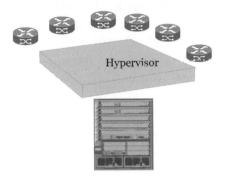

Figure 4.6. *Virtual machines*

Referring once again to Figure 4.6, the basic machine represented here by a router gives rise to six virtual routers. The virtual routers can be connected to one another to form virtual networks, as shown in Figure 4.7.

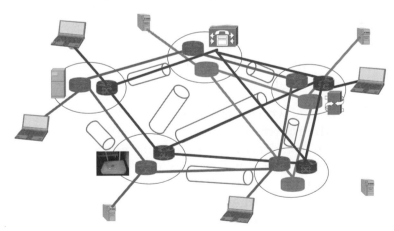

Figure 4.7. *Virtual networks*

Virtual networks are formed from virtual routers using the same protocols. Thus it is possible to create virtual networks with specific protocols – e.g. adapted for VoIP or for IPTV or for any other application. In terrestrial networks, this solution is also widely used to save on energy, always placing the virtual machines at the right place to consume as little energy as possible. For instance, it is easy to change the placement of the virtual routers so that there are fewer physical machines switched on. At nighttime, for example, when the traffic is low, it would be possible to channel all the routes along the same paths so as to turn off as many physical routers as possible. These techniques can be very well adapted to WiFi access points which, by virtualization, would be able to serve numerous operators instead of serving only one. These WiFi access points would be turned off during hours of inactivity, and particularly at night. The difficulty lies in switching them back on at the right time – when a user crops up or to handle an increase in traffic, and being able to guarantee the QoS of the network. There are a variety of strategies which can be employed for this purpose. The most effective of these is to be able to send a wake-up frame to the access point, but this is only possible on the terrestrial network with a hardwired connection. On the wireless network, we have to install a sensor to detect the energy of machines in the process of waking up, either coming into the area or carrying out a handover from a neighboring antenna.

In Figure 4.7, we can also see networking equipment in addition to the routers and antennas. In effect, it is perfectly conceivable to use virtualization to place numerous machines in the existing physical boxes, provided those physical boxes have a high-enough capacity. It is very likely that our Home Gateways will have larger memories in years to come, with increased computation capabilities and a greater number of functions in general. One avenue for research about electrical consumption relates to the repatriation of numerous virtual machines into Home Gateways, which would enable us to put in place distributed Clouds.

4.4. Terrestrial telecommunications networks

Hitherto, we have essentially looked at wireless telecoms networks. We have seen what happens with large telecoms networks for mobiles. In this section, we shall begin by determining where the energy expenditure occurs in terrestrial networks. Figure 4.8 gives a good representation of the situation.

As we can see in Figure 4.8, the energy expenditure is primarily due to the user, slightly less to the local loop and access network, and finally far less to the core network. It is estimated that approximately one Watt per user is spent in the core network, ten Watts on the local loop, and more than 30 Watts on average on the side of the terminal, including the computer or telecommunications equipment. These figures are very logical. Indeed, on the core network, there is a high degree of multiplexing of user flows and networking equipment, although the machines involved inherently consume much more. Indeed, they are shared between a great many users. There is already far less multiplexing on the part connecting the core network to the DSLAM or the equivalent corresponding to the distributor. Multiplexing can be done from anywhere between a few tens to several hundred clients. Such multiplexing is virtually non-existent on the terminal side, meaning up to the plug and the line-end equipment located in the user's home, such as the ADSL router or the Home Gateway. Multiplexing is performed on different traffic streams from domestic users. Consumption is evaluated at around 10 Watts, most of which is attributable to the Home Gateway. Finally, in the user's private space, the home or office network, another stage of multiplexing takes place, but this is usually done by a dedicated user machine. More than the telecommunications part, the electrical consumption is due to the terminal machine which supports the communications protocols.

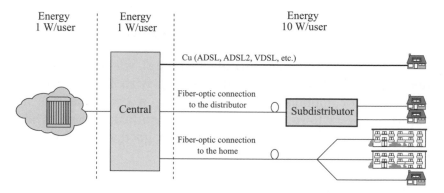

Figure 4.8. *Energy expenditure in terrestrial networks*

We can deduce from all of this that the maximum saving can be made at the level of the local loop, and more so at the level of the end user. The core network is certainly energy-hungry, and improvements can be made to it, but it is actually at the point of the terminal that the maximum gain can be made.

Figure 4.9 is taken from a study presented in the Alcatel-Lucent Research Journal, showing the increase in average electrical consumption per user, which takes account only of technology and adds the results of research into minimizing electrical consumption.

In this figure, we can see that if no progress is made to minimize electrical consumption, the increase is steep, reaching more than 120 W per user. If current research efforts come to fruition and are implemented, the consumption per user should not increase, but rather, in general, should decrease sharply in most cases. This average value should be around 20 Watts per user – six times lower than the value obtained if no progress is made on consumption.

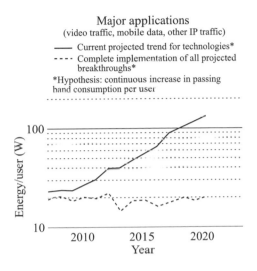

Figure 4.9. *Overall electrical consumption per user*

If we first turn our attention to the core network, current research and development (R&D) is focusing on two points: how to make networking equipment energy-efficient and how best to multiplex the resources so as to turn off as many machines as possible.

In the first direction of research, the question is how to develop processors capable of adapting to the workload: the less work that needs to be done, the more the processor should slow down, even to the point of stopping when there is no traffic to deal with. In that same direction, research is also being put into electronic circuits which consume less and less energy. In terms of progress within a short time period, this is undoubtedly the avenue which offers the greatest possibilities.

A second avenue of research, mentioned above, relates to multiplexing routes along shared paths so as to switch off as many machines as possible, or at least to allow the network equipment to enter sleep mode without actually shutting down. For instance, at night, the networks are totally under-used, and it might be advantageous to put as many machines on standby as possible while maintaining an acceptable level of QoS. Virtualization, which was discussed earlier on, constitutes the main technology to be implemented, grouping the active virtual machines together on shared nodes. Of course, the connectivity constraints are significant, and all the active users must be able to be connected.

As regards this latter solution, work is currently being done on utility functions for forwarding nodes. The utility function must be equal to zero in order for the machine to stop. This utility function may take account of the connectivity of the other nodes, and of course, of the traffic coming through the node in question.

On the side of the local loop, energy-saving solutions are more complex, but the gains made by implementing them may prove far greater. The first solution, chronologically, relates to the installation of fiber optics which, thanks to technologies such as PONs (Passive Optical Networks), is able to multiplex 50 users on a single optical fiber. The PON access diagram is shown in Figure 4.10. The technical solution of a PON is a standardized technology, and may come in the form of an EPON (Ethernet PON) or a GPON (Gigabit PON) which also uses Ethernet frames in a framework standardized by the ITU-T.

Another significant opportunity relates to the consumption of communications boxes, whether these are "Home Gateways", TV decoders or WiFi access points. All three boxes consume around 10 W of power, the decoders consuming the most, followed by the Home Gateways and then the WiFi access points. The combination of the three pieces of equipment reaches up to between 20 and 30 W. New technologies should be able to markedly reduce this consumption over the course of the next few years, taking it down to less than 5 W for all three and 2 W for a single WiFi access point. This decrease, when multiplied by millions of boxes, represents a substantial energy saving. On the other hand, new generations of WiFi could increase consumption, e.g. the IEEE 802.11n.

Another way to economize on these boxes would be to put them on standby. Today, this solution is not automatically implemented, and users themselves have to turn off their boxes. With the current state of affairs, even in a non-functioning regime when the PC connected to it is switched off, the box's electrical consumption is practically the same. Stop and Start techniques ought to be able to be used to put the box on standby when its utility function falls to zero. Obviously, the telephone line has to remain active even when the box is in sleep mode.

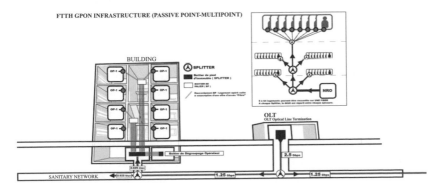

Figure 4.10. *A PON (Passive Optical Network) access point*

Unfortunately, WiFi access points and cards have been designed in a rather similar way: it is near-impossible to put them in sleep mode without switching them off completely. Yet if we look at a typical domestic user, he only uses his

Internet connection a few hours a day at most – sometimes less than an hour. Contemporary access points often consume far more than 10 W, but this figure can be brought down to 5 or 6 W in rather exceptional cases. Consumption is fairly closely linked to the power of the WiFi signal, which in most countries is itself subject to regulations, and cannot be greater than 100 mW. In fact, portable PCs often have WiFi emission strengths of around 30 mW, and smartphones often transmit at strengths below 10 mW.

It should be noted that IEEE 802.11n access points, and even more so the new generations IEEE 802.11ac and af, consume a great deal more energy, between 20 and 50 Watts.

The terminal machines with which the user interacts consume far more energy than the telecoms boxes do. Mobile telephones connected to the commuted telephone network consume around 6 W, modem-routers around 12 W, VoIP modems around 10 W, and desktop computers can consume up to 200 W in a situation of very intensive usage, and 50 W in idle mode, i.e. when there are no tasks being performed on its system. The power expended depends on the number of output and graphic cards, on the memory, on the power of the processor, the cooling system, etc. To give some illustrative examples of the orders of magnitude, Ethernet cards consume between 3 and 4 W, graphics cards up to 50 W, PCI cards between 5 and 10 W, a stripped motherboard between 20 and 40 W, a live memory between 5 and 8 W. We can immediately see that a great deal of energy could be saved by turning off those components which are not in use, and installing low-consumption components. One of the most promising points from this perspective relates to screens (known as monitors), which can consume up to 150 W. It should be noted that even putting the monitor on standby does not completely eliminate its consumption, which remains between 1 and 2 W.

A far better prepared extremity machine can be seen in portable computers (laptops), which consume around 25 W in a typical configuration. This value is sufficient for modern-day batteries to last for several hours' worth of work.

Rather than completely shutting down a network machine or a terminal machine, the solution which is increasingly used is to shut down parts of those machines, one component at a time. For instance, a PC might leave only its Ethernet card running, in order to receive data and restart the system when a packet is received.

4.5. Low-cost and energy-efficient networks

We have seen that the access networks are one of the most significant sources of energy consumption. An idea that is beginning to be widely adopted is to have low-cost access networks with low energy consumption. Indeed, given that the consumption is largely dependent on the distance squared, we have to minimize distances between terminals and therefore increase the number of access points. To this end, two solutions would seem to suggest themselves: femtocells and mesh networks.

Let us begin by comparing mesh and *ad hoc* networks. These are networks such as those shown in Figure 4.11. Mesh networks are infrastructural networks because they require machines belonging to the network operator. Another solution is also feasible: *ad hoc* networks, which are also illustrated in Figure 4.11. *Ad hoc* networks are made up of machines belonging to the end user, and the routing software is on the user's machine. This solution is more complex in terms of the heterogeneity of the machines, and of management and control of them. *Ad hoc* networks can also be used in fairly simple cases, because – due to the movement of the clients themselves – it is very difficult to ensure good QoS.

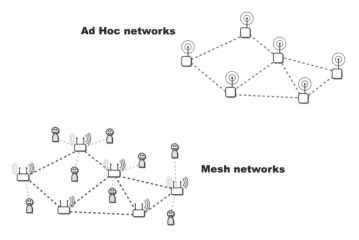

Figure 4.11. *Mesh and ad hoc networks*

Mesh networks represent an excellent solution for access networks in developing countries, and for extending networks in developed countries. One of the advantages to such networks relates to the total cost of the network, because for the same QoS as that obtained using a 3G antenna, firstly the expenditure is 20 times less for the mesh network, and secondly, the energy expenditure is a 100 times less, in comparison to what can be achieved using 3G antennas.

In the context of mesh networks, the client is connected to the mesh access points, and the traffic in packets is passed from node-to-node until it reaches the addressee or the Internet access. The nodes forming the network must not be positioned too far from one another so as to ensure good QoS. The consumption of a node can drop to less than 10 W. Around 20 mesh access points can yield the same overall data rate as a 3G antenna.

Therefore, energy savings relate firstly to the actual consumption of the machines involved. A second avenue which is also being extensively explored is that of overlapping paths so as to optimize the number of machines that can be turned off. An example of this solution is shown in Figure 4.12. The three terminal machines are connected to the three access points to begin with. Then, the routing is modified so as to use the same WiFi access point. Consequently, the four machines which are no longer being used to interconnect the users can be put in sleep mode.

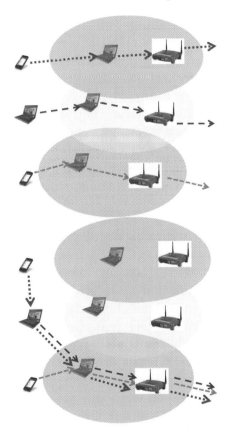

Figure 4.12. *Optimization of the paths*

Another avenue of research relates to femtocells. At the center of the femtocell is the Home Node-B (HNB) – generally a card inserted into the Home Gateway, with an antenna to link mobile equipment such as smartphones. This connection can be made using a 3G or 4G frequency offered by the operator associated with the femtocells. However, the tendency which we are seeing today is to make this connection using a WiFi environment. Indeed, the cost of 3G/4G frequencies is particularly high, and the explosion in WiFi offers a far less cumbersome solution.

One solution for handling data flows is to create a mesh network of Home Node-B's (HNBs), enabling the packets to be sent out via the fiber optic or ADSL connections available one or a few jumps from the transmitter. We have illustrated this type of solution in Figure 4.13. This figure shows that an operator can introduce Home Gateways with no direct connection to the core network, connecting them via the mesh network to a fiber optic connection located one or a few jumps away. Start and Stop technology can automatically switch off Internet boxes when their utility function drops to zero.

Another solution which is developing rapidly and which, fundamentally, is not too different from the previous one, is NGH (Next Generation Hotspot) access points. These are WiFi access points handled by telecoms operators and which behave like 3G/4G antennas. The advantage to this solution from the perspective of electrical consumption involves the same arguments as before: the shorter the distance between the user and the access point, and the less power required, the more NGH access points can serve as relays, and finally the nodes can be equipped with the Start and Stop function, to greatly economize on the energy consumed. These access points conform to the standard IEEE 802.11u, the aim of which is to facilitate a connection which is entirely similar to that made using a 3G/4G antenna. Consequently, from the point of view of the user, there would no longer be any difference between a WiFi access point and an antenna of a mobile telecommunications network.

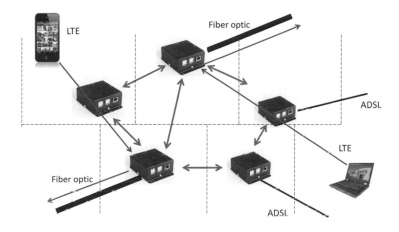

Figure 4.13. *A mesh network of HNBs*

4.6. The role of virtualization in "green" techniques

We have already touched on the importance of virtualization in the next generations of telecoms networks, and on some elements of its effect in terms of energy savings. In this section, we shall come back to these elements, and go into greater detail about different aspects. Let us recall, first of all, that virtualization creates a software version of what was previously a piece of hardware. The functions served by the hardware are described in the form of software, which must be executed on a machine powerful enough to match the performances of the former hardware. The limitations of virtualization are found in extreme cases, where the power of the physical machine can no longer be matched by the equivalent software package. With the dawn of data centers on a massive scale, this limitation in terms of performance has been pushed back to a large extent.

The advantage to virtualization lies in the facility of adding a new virtual machine to a physical machine. If that physical machine is very powerful, numerous virtual machines can be hosted on it, which can be started and stopped easily. We can load new virtual machines onto the physical one, as well as take old ones off it. As already pointed out, this technology enables virtual networks to be created, by linking the virtual machines of the same operator, e.g. an IPv4 network, an IPv6 network, an MPLS network, etc.

The gain in terms of energy saving relates largely to the possibility of sharing a common physical infrastructure. For instance, a WiFi access point can be shared between several operators. Virtualization may be viewed in two different ways: the classic case where the access point has several SSIDs – i.e. several names – and the case of virtualization of networks, whereby the virtual networks are totally different. These two cases are illustrated in Figure 4.14: network virtualization above and VLAN virtualization below.

Network virtualization enables networks to be constructed whose protocol stacks are entirely dictated by the operator, to conform with that operator's core network. Conversely, in the latter case, the protocols are the same between the different networks, which are not isolated from one another.

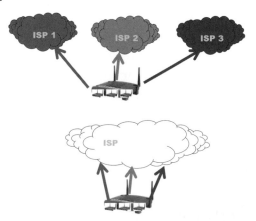

Figure 4.14. *Virtualization of a WiFi access point*

Very substantial energy savings can be made by optimizing the number of physical machines and grouping virtual machines together on shared physical platforms. Data centers can be turned into generic physical machines by grouping the virtual machines together. The energy saving is then pushed to a new level: that of servers which can be turned off in the data center. An even higher level would be to group the virtual machines together across a few data centers so as to turn off as many data centers as possible. Standards issued by the Internet Engineering Task Force (IETF) suggest two rather different protocols in order to carry such shifting of virtual machines: TRILL and LISP.

TRILL (TRansparent Interconnection of Lots of Links) enables packets to be transferred between two machines with no preliminary configuration; the transfer is entirely secure, even during periods when there might be loops, and the procedure supports both unicast and multicast protocols. TRILL performs this task using IS-IS (Intermediate System-to-Intermediate System) as a stative routing of the links, and encapsulating the traffic by using a header which contains a number of jumps. The machines used by TRILL are called RBridges (router-bridges). TRILL supports LAN (Local Area Network) multi-access – access links which can have several terminal stations connected to RBridges.

Virtual machines can also be transported from one data center to another. The IETF advocates LISP (*Locator/ID Separation Protocol*) to handle these transfers of huge quantities of data.

The basic idea behind locator/ID separation relates to the architecture of the Internet, which combines these two functions: the routing locator (the network attachment) and the ID ("Who is it?"). The proponents of separation architecture postulate that the separation of these two functions offers numerous advantages, including greatly-improved evolutivity for routing. The aim of this partition is to improve the efficiency aggregation of routing space locators and to provide persistent identifiers in identity space. LISP also supports different IPv4 and IPv6 address spaces. It should be noted, however, that in LISP, the IDs and locators are in fact IP addresses. The IDs comprise two fields: a "global" and unique field, which identifies an interface in a site, and a "local" field, which identifies a particular interface within the site. The "local" field may be split into two so as to identify a particular network in the site as well. For a given identifier, LISP correlates the "global" field of the identifier with a set of locators that can be used by an encapsulation to get the identity of the interface. The result of this is that a host can change its identifiers when it moves from one site to another, or each time it moves from one sub-network to another within the same site. As a general rule, the same IP address will not be used as an identifier and as a locator in the same LISP environment.

We can see that the modern world of networks will be totally virtualized, with virtual machines and more specifically, virtual resources, which will be optimally positioned so as to switch off as many servers (or data centers) as possible.

4.7. Conclusion

Telecommunications networks require a great deal of energy to function and their consumption burgeons rapidly as data rates increase. If we are not careful, their proportion of the carbon footprint could grow from 2% to 20% within 10 years. Fortunately, numerous solutions are being implemented in order to attempt to prevent too large an increase. Among the solutions which we have discussed are: virtualization techniques, which enable more appropriate multiplexing and the switching off of useless physical resources; low-consumption access networks, whose energy demands are a great deal less than such requirements typically stand at the moment; and network cards which can be uncoupled from the central processing units (CPUs) so as to be switched off when necessary.

We can anticipate numerous breakthroughs yet to come, far more significant in terms of impact than inventions up until now. Of the more or less advanced techniques, one might cite: data- and packet-processors whose speeds can adapt to the workload to be performed; new techniques for sometimes transmitting outside of normal radio frequencies to achieve far greater data rates and consume far less; even higher degrees of multiplexing; low-consumption memories; heat recovery to recycle energy, and so on.

4.8. Bibliography

[GOM 06] GOMEZ J., CAMPBELL A.T., "Variable-range transmission power control in wireless ad hoc networks", *IEEE Trans Mobile Comput 6*, p. 87-99, 2006.

[GUP 02] GUPTA P., KUMAR P.R., "Critical power for asymptotic connectivity", *37th IEEE Conference on Decision and Control*, p. 1106-1110, 2002.

[GUP 07] GUPTA R., MUSACCHIO J., WALRAND J., "Sufficient rate constraints for QoS flows in ad-hoc networks", *Ad Hoc Networks 5*, p. 429-443, 2007.

[MAR 12] MARTIN S., AL AGHA K., PUJOLLE G., "Traffic-based topology control algorithm for energy savings in multi-hop wireless networks", *Annals of Telecommunications*, vol. 67(3-4): 181-189, 2012.

[ODO 09] ODOU S., MARTIN S., AL AGHA K., "Admission control based on dynamic rate constraints in multi-hop networks", *IEEE Conference on Wireless Communications & Networking Conference (WCNC)*, p. 1956-1961, 2009.

PART 2

A Step Towards Smart Green Networks and Sustainable Terminals

Chapter 5

Cognitive Radio in the Service of Green Communication and Networking

5.1. Introduction

Cognitive radio networks [AKY 06] are emerging as a new concept for communication and management of increasingly scant wireless resources. The objective is to exploit residual bandwidths in the frequency spectrum. Indeed, a number of recent studies [FCC 02] have highlighted the disparity and sub-optimal use of the radio spectrum, stressing the fact that certain bands – particularly the free bands – are becoming overloaded, whereas others remain largely under-used. In this context, cognitive radio has been advanced as a technology capable of exploiting the residual bands during periods when their licensed users (known as primary users) are not broadcasting on those frequencies, and to free up the channel as soon as the primary users attempt to access it. In order to do this, cognitive radio relies on the concept of software defined radio (SDR), capable of dynamically

Chapter written by Hicham KHALIFÉ.

switching to a specific channel in the spectrum in real time, to transmit on that channel or "sense" its activity. Cognitive radio technology adds to this software layer a dimension of intelligence, learning and adaptation (Figure 5.1). This renders radio equipment capable of predicting the usage of the spectrum and making appropriate decisions in real-time.

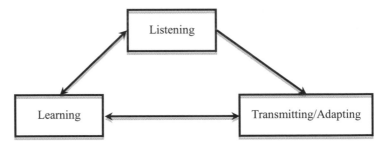

Figure 5.1. *How cognitive radio works*

Thanks to this property of adaptation and dynamic selection of frequencies, cognitive radio can be exploited to make wireless equipment more energy efficient. Today, it is clear that extending the concept of green energy to communications networks is becoming an inescapable necessity. According to recent studies, 3% of the world's energy consumption is invested in information and communication technology (ICT) infrastructures. Solutions are needed in the medium-term to contribute to the reduction in the amount of greenhouse gases produced by the domain of ICT, estimated to be roughly 2% of global emissions. This observed level constitutes a similar percentage as the CO_2 produced by air transport. With the aim of reducing the energy consumption of communication technology while maintaining the same QoS, new transmission techniques, new technologies and optimized protocols must be put forward. In this context, wireless access technologies – which are becoming increasingly widely used because of the rapid expansion of WiFi and 3G/4G – constitute a main source of this consumption. This is manifest in the mass installation of infrastructures for such networks, as well as in the usage of the batteries of communications equipment such as smartphones. Thus, it is clear that optimization of the deployment and harmonization of the function of the different infrastructures are interesting avenues to explore in order to ensure "green" communications. In addition, in order to prolong the battery life and reduce recharging cycles, it is necessary to optimize or coordinate the functioning of the wireless interfaces used by such equipment. Cognitive radio offers the flexibility to cater for these needs.

Various initiatives to do with green energy in networks have already been proposed in the form of consortiums, research actions and collaborative projects. The best known of these initiatives is GreenTouch [GRE b], the aim of which is to increase the energy efficiency of ICTs a thousand fold. The consortium has set 2015 as a deadline to put forward an energy-efficient communication architecture to demonstrate the gains that can be made. Other initiatives which focus on wireless networks and more recently on cognitive radio ([COG] and [GRE a]) have been advanced. The goal of these initiatives is to analyze, define and optimize energy consumption in wireless networks and communications. However, these works, and particularly the use of cognitive radio for green purposes, are still at a preliminary stage. In parallel, a number of papers have been published, providing directives for green techniques in cognitive radio [PAL 09; GÜR 11]. In this chapter, we give an overview of the main advances in the field and propose future avenues for research.

The rest of the chapter is organized as follows: in section 5.2, we detail the concept of cognitive radio and present a number of projects and initiatives in that domain. We give different definitions of "green" in cognitive radio in section 5.3, and then put forward a variety of green solutions using cognitive radio in section 5.4. A concrete use case which exploits these solutions is presented in section 5.5. We conclude and highlight our vision of the future in section 5.6.

5.2. Cognitive radio: concept and standards

Unlike other transmission media such as cables or fiber optics, the wireless support is difficult to extend. Indeed, the data rate that can be obtained by way of wireless transmission is theoretically bounded. The capacity of hardwired supports is easy to alter, by laying as many cables as are needed. The same is not true for wireless transmissions which, even with the most high-performance encoding techniques, still offer a limited data rate because of well-defined theoretical boundaries.

Faced with the saturation of free frequency bands, but more recently of 3G bands as well, cumulating the bandwidth onto other frequencies appears to be the only way of increasing the data rate. This observation sparked research into cognitive radio in light of the FCC report in 2002. Since then, considerable effort has been invested, and numerous proposals have been made. Nevertheless, before this technology comes into its own and can be released to the wider public, certain hurdles still need to be overcome. To begin with, cohabitation between secondary nodes, which will exploit the available frequencies to cumulate the bandwidth, and the primary nodes which have priority over the channels, constitutes a genuine problem for operators. Obviously, operators were opposed to sharing the bands they had bought or sold with opportunistic users in the absence of any real guarantees. Things began to move forward rapidly once the operators themselves found they were facing problems with data rates on already-allocated bands. Another problem lies in defining the economic model which would launch this new technology. The actors in this field are searching for a model which would benefit both users and operators at once.

5.2.1. *Attempts at standardization*

Numerous attempts at standardization accompanied the emergence of cognitive radio. Several organizations have put forward norms and standards; other initiatives are still at the stage of preliminary examination. Below is a summary of the most advanced works:

– the *ITU World Radiocommunication Conference (WRC)*, organized every three years, is looking into regulating the frequency spectrum in view of the new technologies and applications. The next session, programmed for 2012, will focus on software-defined radio and cognitive radio in light of the results of the ITU-R study;

– the *FCC* has promised to find 500 MHz of additional spectrum width for high data rate communications. To date, only 25 MHz have been allocated. However, the FCC continues to defend a more flexible use of the spectrum, without primary users having higher priority to access their own dedicated frequencies;

– the standard *IEEE 802.22 WRAN* constitutes the single most successful initiative for standardization in the context of cognitive radio. The standard suggests using analog and digital TV bands and the bands used by microphones to form regional networks which can cover around 100 km. The standard published in July 2011 puts forward a framework for the use of cognitive radio in infrastructure mode in regional networks;

– *IEEE SCC 41 (formerly IEEE 1900)* is a series of standards developed by the standardization committee on Dynamic Spectrum Access Networks (DySPAN). The aim of this initiative is to improve the usage of the spectrum, control the level of interference, and to optimize and coordinate the different wireless technologies, including information management and sharing. The committee is structured into seven working groups, each one focusing on a specific aspect of this technology;

– the standard *IEEE 802.11af*, also known as White-Fi, has recently been put forward to allow the standard IEEE 802.11 (WiFi) to operate in TV white spaces. The standard proposes to use geographical databases to centralize information about the availability of bands. Thus, the databases are consulted before each transmission. This new standard promises higher data rates owing to transmission on wider bands. This is made possible by using techniques which allow transmission on non-contiguous channels.

There are other attempts at standardization, but we shall not mention them in this chapter because of their low impact.

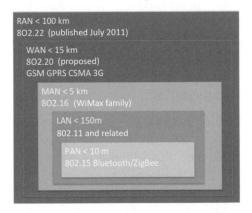

Figure 5.2. *IEEE wireless standards*

5.2.2. *Research projects and initiatives*

Numerous projects, largely financed by the European Union, have looked at cognitive radio and attempted to offer economic models as well as prototypes and platforms for validation. Below, we summarize the most pertinent of these initiatives:

– *LICoRNe (2010-2013)*: financed by the Agence Nationale de la Recherche (ANR) in France, the LICoRNe project is investigating the engineering of the different services to offer users in a multi-hop cognitive radio network. The project proposes to validate the solutions it advances by testing them on a cognitive radio platform based on GNU radio. In addition, specific scenarios will be tested on the network of a wireless operator;

– *SENDORA (2008-2010)*: this is a European project (ICT-FP7) that aims to use special sensors to detect opportunities for transmission in the radio spectrum. The purpose of these sensors is to help the cognitive radio nodes to detect available bands so as to avoid any interference with the primary networks;

– *CREW (2010-2012)*: the goal of this European project (ICT-FP7) is to set up an open cognitive radio test platform. In this project, five open platforms will be deployed, each managed by a partner and able to host experiments. CREW is intending to use USRP GNU radio devices on these five platforms, linked to one another via the Internet;

– *QASAR (2010-2012)*: this is a European project (ICT-FP7) which aims to reduce the gap between the theoretical studies on cognitive radio and its real-world implementations. One of the goals of the project is to study the economic model; another is to investigate regulations in the domain;

– *CROWN (2009-2012)*: this European project is intended to make cognitive radio networks viable both technologically and financially, by putting forward techniques for efficient use of the spectrum. This project proposes to build a demonstrative prototype to validate its approach. One of the major assets of the project is that Ofcom is one of its partners; Ofcom is the organization that regulates the spectrum in the UK;

– *QoSMoS (2010-2013)*: the aim here is to develop a cognitive radio prototype in order to design a true product. The partnership envisages testing their prototype with applications in TV white spaces;

– *SACRA (2010-2012)*: this is another cognitive radio project intended to design cognitive radio equipment and algorithms which would facilitate communication on several radio frequencies simultaneously;

– *FARAMIR (2010-2012)*: this project aims to characterize the use of the spectrum by way of measuring campaigns. Thus, its focus is on the techniques needed to carry out large-scale measurements, then on sampling and on the exploitation of the vast quantities of data thus obtained;

– *C2POWER (2010-2012)*: this is another European project which intends to exploit cognitive radio with the goal of reducing energy consumption in wireless networks. The main idea is to exploit collaboration between different cognitive radio users so as to reduce energy consumption. Such collaboration helps gather information about which bands to use and the transmission power on each of the available frequencies. Finding incentives to persuade users to collaborate and share the information needed to optimize energy consumption constitutes a major axis for the work of this project.

Other projects such as CogEU propose to exploit software radio and the concept of cognitive radio, but within the limited context of TV white spaces.

5.3. Various definitions of green in cognitive radio

A number of dissimilar definitions exist today to describe what "green" is, in the context of cognitive radio. This is mainly because the domain is so new; a certain degree of maturity is still needed before consensus can be reached and a consistent body of terminology drawn up. More specifically, the three definitions of cognitive radio all deal with a specific parameter affected by the functioning of cognitive radio devices: that of dynamically choosing the channel on which to transmit. It is important to note that some of these definitions are not entirely incompatible and could be brought together. They essentially relate to a direct or indirect action as regards the transmission power employed by cognitive radio equipment.

5.3.1. *Reducing the pollution of the radio spectrum*

Favored by the expansion of the use of wireless equipment, the noise recorded on many frequencies is now at record levels. There is a clear and present danger that this pollution will reach certain critical bands. Because of its dynamic use of channels, cognitive radio can reduce the pollution on certain critical bands by choosing to transmit on less busy frequencies. This intelligent technology could even maintain a security margin around the most important channels. This is illustrated in Figure 5.3.

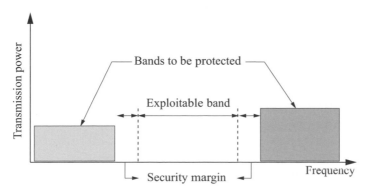

Figure 5.3. *Thanks to cognitive radio, it is possible to safeguard sensitive bands*

5.3.2. *Reducing the exposure of individuals*

Another of the objectives of green in wireless networks may be to control the level of exposure of individuals to electromagnetic waves. The debate about the impact of wireless transmissions and waves on people's health has already been raging for many years. This manifests itself by opposition to the installation of new transmission antennas in urban areas. By facilitating dynamic adaptations of the bands used and on-the-fly alteration of the parameters of communications (such as the transmission power for instance), cognitive radio can limit the degree of exposure to individuals, in certain areas, to radio waves.

5.3.3. *Reducing the consumption of the equipment*

This definition corresponds to the conventional idea of green technology that is widespread today. Indeed, reducing the energy consumption of the machines by optimizing their mode of function or even by turning them off, helps increase their lifespans and prolong their recharging cycles. In particular, cognitive radio offers control over the transmission power on each of the wavelengths chosen. It is clear that the lower the transmission power is, the lower the energy consumption will be. However, if we reduce the transmission power, both the data rate and the coverage radius are adversely affected. By choosing the correct frequency band, cognitive radio technology can optimize these three parameters with the aim of improving energy consumption. These compromises will be explained in detail further on in this chapter.

5.4. Clean solutions offered by cognitive radio

In this section, we discuss the cognitive radio solutions which can be harnessed in the service of green communications. We also highlight ongoing initiatives and research projects relating to each of these solutions.

5.4.1. *Solutions for the spectrum and health*

Not a great deal of work has been devoted to the aspect of the user's health of green technology in cognitive radio. In particular, here, we wish to cite the works of Palicot [MIC 10; PAL 09] which stress the harmful effect of the use of omnidirectional antennas in wireless terminals. This exposes the body and particularly the head to high quantities of electromagnetic radiation. The ideas and suggestions proposed in this paper to reduce exposure to radiation notably involve the use of directional antennas on the bands being used. Indeed, it seems there is a need for dynamic configuration of the antennas, the transmission powers and other physical parameters of the signals emitted. However, such adaptations – while technically feasible with cognitive radio – may be very difficult to implement in the case of a moving user and/or changing properties of the transmission channels.

5.4.2. *Actions at the level of equipment/infrastructure*

A considerable reduction in energy consumption can be achieved by acting directly on the communication infrastructures. It is our belief that cognitive radio allows us to limit the number of active base stations but also to reconfigure them so as to exploit specific frequencies. For instance, in certain circumstances, it may be advantageous to keep only one piece of wireless equipment active in an area, and switch off all the other access points. In this context, the terminals (equipped with cognitive radio technology and therefore capable of using different radio bands) wishing to communicate will use the available band. However, an increase in radio activity will require other access points in the area covered to be brought into play. In order to reduce the workload and strain on the existing infrastructure, it is clear that these access points will have to operate on frequency bands other than those already active in the network. This is illustrated by Figures 5.4 and 5.5.

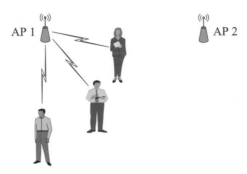

Figure 5.4. *Access point active (AP1); all three users are using the same band*

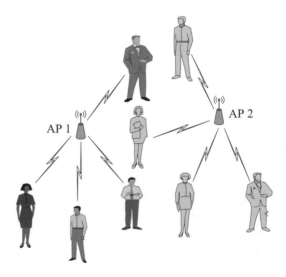

Figure 5.5. *New users appear; access point AP2 begins transmitting on another frequency band*

Today, approaches to act on wireless communication infrastructure have been put forward in the context of research projects. The EARTH Project [EARTH] is the best-known initiative in this field. Specifically, this project aims to adapt wireless infrastructure with the goal of reducing its energy consumption. In order to do so, it proposes to exploit new technologies such as MIMO, but also coordination between the different terminals, power control and (re)division of the cells. EARTH does not explicitly cite cognitive radio as a technology capable of acting upon and reducing the energy consumption of communication infrastructures.

5.4.3. *Optimizing the communication parameters*

On a shorter timescale, it is also possible to reduce energy consumption by intervening in real time on the end user terminal. This can be done using cognitive radio protocols capable of adapting the communication parameters on the fly. Below, we detail the key parameters which, when optimized, are able to reduce the energy consumption of the equipment.

5.4.3.1. *Bandwidth and capacity*

According to the well-known Shannon-Hartley theorem, the capacity (in bits/s) obtained on a transmission channel depends on the bandwidth W (in Hz), but also on the signal-to-noise ratio (S/N) on the transmission link:

$$C = W \log_2(1 + S/N)$$

This formula clearly shows that the capacity increases in a linear fashion with the bandwidth (W) but only logarithmically with the transmission power. Indeed, the higher the transmission power, the better the signal-to-noise ratio, but higher transmission power also means greater consumption of the battery life. This observation demonstrates that in order to increase the data rate it is by far preferable to increase the bandwidth; thereby, the increase in data rate will be larger for the same level of energy consumption. Consequently, for the same data rate it is possible to limit the energy consumption by reducing the transmission power but increasing the bandwidth at the same time.

Grace *et al.* [GRA 09] go further. More specifically, they deduce a direct relation between the battery life of the mobile terminal, the bandwidth and the data rate. The authors also show that the transmission power of a base transceiver station can have a negative impact on the energy consumption of the terminals on a nearby cell. Logically, the higher the transmission power of the access point, the more interference (noise) will be generated for non-addressees, and the more the terminals on neighboring cells will have to increase their transmission power so as to maintain the same signal-to-noise ratio. This causes the terminals to have limited lifespans. Cognitive radio, by exploiting different frequency bands in the cells, is able to temper this effect.

5.4.3.2. *Tradeoff between bandwidth, transmission power, distance and data rate*

In order to reduce energy consumption, cognitive radio has to optimize a complex tradeoff between bandwidth, transmission power and capacity (data rate). As explained above, if the transmission power is reduced, the data rate of the communication also falls. Although this can be compensated by (slightly) enlarging the bandwidth of the channel, on the majority of frequency bands used, the opportunity to do so is non-existent. In addition, it is well known that transmissions on low-frequency channels cover greater distances because of their resistance to attenuation and to interference. In concrete terms, if two nodes placed far apart are not within range of each other using a given frequency, it is possible by using a lower frequency *and with the same transmission power* that these two terminals may be able to communicate. Conversely, the data rate of that communication will certainly be less, because if one transmits on a lower frequency, the bandwidth is also reduced.

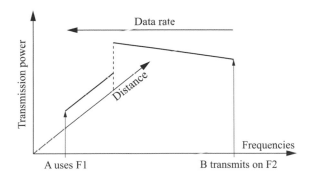

Figure 5.6. *Compromise and cognitive radio*

Thus, it proves possible to ensure "green" communications using cognitive radio – a technology that is capable of dynamically adapting all the parameters involved in a transmission. The different parameters are represented in Figure 5.6. This diagram shows that a cognitive radio terminal A, by choosing a lower frequency (F1) and transmitting with less power, can achieve communication across distances equivalent to another terminal B which transmits at a higher power on frequency F2. This illustrates the fact that cognitive radio, by dynamically choosing lower bands, can increase the lifespan of the machinery and create a greener network. However, this gain in lifespan comes at the expense of the data rate obtained. This constraint is not an obstacle when the lowest frequency band chosen ensures a satisfactory data rate for communication. However, in certain cases, the data rate required cannot be guaranteed using lower frequencies (see Shannon's formula). It is clear that work aimed at ensuring higher data rates by optimizing the modulation, encoding, antennas and other physical parameters for the communication may prove useful in the future.

5.4.4. *Avenues for research and visions for the future*

Given that research on green communications using cognitive radio is still in its infancy, below we present a number of perspectives and ideas to be exploited for the future. In practical terms, these solutions consist of combining several of the unitary actions detailed above so as to offer a complete solution.

5.4.4.1. *Exploiting multi-hops and mobility*

A reduction in the transmission power of certain machines, and thereby their energy consumption, can be achieved by exploiting multi-hop communications. In this context, intermediary nodes can play the role of relays, thereby increasing the range of the transmitters without having to increase the transmission power. These kinds of solutions could be delivered by cognitive radio to increase the lifespan of the network. This strategy can be made even more effective if collaborative techniques between the different nodes (terminals and infrastructure) are put in place. Collaboration enables us to gain from experience, from vision and from previous communications with nearby nodes to reduce energy consumption. However, in a multi-hop network, green solutions can also be used at the level of routing and the pathways taken by the messages.

Future solutions must also exploit the mobility of users as a factor which reduces their energy consumption. In practical terms, two types of mobility exist in the use of cognitive radio:

1) physical mobility on the part of the users; and

2) mobility (availability) of the exploitable frequency bands.

These two types of mobility define the neighborhood and the topology observed by a cognitive radio node. In other words, two nodes are considered to be neighbors if they are physically close and if they can use at least one common frequency band. These two mobilities, used positively, can help reduce energy consumption. For instance, if an access point is aware of the mobility of the intended recipient, it can send him a message when he comes closer physically or passes into a low frequency band. This enables the emitter to transmit at a lower power, and therefore decrease its energy footprint.

5.4.4.2. *Cross-layer solutions, bringing scientific communities together*

Like in all wireless networks, cross-layer solutions are needed to ensure green communications in cognitive radio. Indeed, these solutions enable us to act on several layers of the protocol stack at once, and to simultaneously optimize several parameters. However, the high number of adaptable parameters in cognitive radio renders the cross-layer approach even more complex.

In particular, in order to put forward green solutions, any initiative must be able to simultaneously:

– act directly on the machines, to switch them on or put them on standby;

– alter the parameters of the physical layer, such as the chosen frequency, the bandwidth, the transmission power, the modulation, etc., "on the fly";

– adapt the channel access protocol depending on the frequency being used. Indeed, the performances of the MAC layer are also related to the properties of the underlying channel;

– make use of virtualization techniques to change the role or the mode of function of a piece of equipment. Thus, a base station can be reprogrammed to use a different technology if the frequency bands which that new technology exploits facilitate a reduction in energy consumption.

Figure 5.7 shows the range of cross-layer solutions to offer green consumption using cognitive radio.

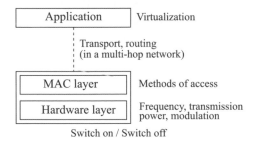

Figure 5.7. *Cross-layer approach to green techniques*

Recently, the C2Power ICT/FP7 project [C2Pow] has been attempting to use cognitive radio in a "full cross-layer" approach, with green constraints to satisfy in each layer. Because this promising project is still in its formative stages, concrete solutions have not yet been put forward.

5.4.4.3. *Thinking green*

Our vision of the domain consists of saying that, with the aim of finding efficient cognitive radio solutions whose complexity is not untenable, we have to "think green". In other words, the solutions proposed must, at their very heart, from their conception, be conceived for green energy, rather than being an adaptation of existing techniques. This would enable us to consider green energy as a main constraint and avoid it becoming a secondary constraint added into the system. Needless to say, this vision has to be global, and necessarily cross-layer.

These efforts must also take account of the inter-community aspect of cognitive radio technology. Indeed, because it necessitates cross-layer solutions, cognitive radio is a domain which simultaneously affects several research communities. Skills in signal processing, in computing and sometimes in physics are needed to offer green solutions. This new field of research should in time forge connections between these different communities in the service of the environment.

5.5. Use case: "Smart buildings"

The buildings of the future will certainly include an increasing number of technological devices. These devices will consume a large amount of energy. In the field of construction of high environmental quality (HEQ) housing, it is also necessary to reduce the consumption of communication devices. This can be done by setting up collaboration between the different technological components, and introducing solutions which have become possible thanks to cognitive radio, with the goal of optimizing the function of the communications infrastructure and limiting the inhabitants' exposure to radio waves. Below, we present a number of avenues for research which we find particularly noteworthy:

-- In a smart building, the communications infrastructure will use cognitive radio technology to put itself on standby dynamically. Only one service will remain active, probably on the frequency which consumes the least energy. Thus, as detailed above, the users synchronize on the available band in order to communicate. Direct action on the equipment making up the wireless infrastructure has become possible thanks to recent advances in virtualization techniques. In parallel, femtocell technology – which consists of reducing the transmission power of GSM/3G base transceiver stations (BTSs) by installing mini-BTSs directly in the user's home –

seems a good candidate for achieving green energy usage in smart houses. It thus becomes clear that the fact of being able to simulate several machines on a single piece of hardware means that a local GSM station (a femtocell) can reconfigure itself as a WiFi access point in sleep mode, for instance. Clearly, the location of the access points, the frequencies used and the turning on or off of the equipment must be optimized in view of the position and mobility of the users and their terminals.

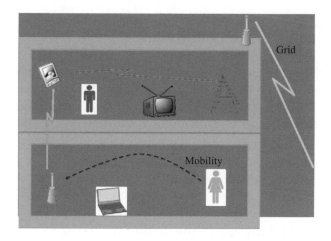

Figure 5.8. *Cognitive radio for a smart building*

– In communications, terminals will be able to dynamically adjust their communication parameters so as to reduce energy consumption. Indeed, cognitive radio is able to change the frequency band used, and modify the modulation and the transmission power in order to make energy savings. This involves optimizing the data rate/bandwidth/distance tradeoff. This seemingly simple operation may prove complex when numerous terminals and access points have to be configured dynamically, on different floors of the building. Since we know that local configurations can have an impact on neighboring communications on the same frequencies, rapid and powerful algorithms have to be put in place.

– Within the building, the mobility of the residents should be exploited to make communications green. By recording the patterns of movement of the different individuals, it becomes possible to predict their behavior. When intelligently exploited, such a prediction enables an arrangement to be set up at the level of the access points to favor accessible users with low energy consumption. For instance, an access point would give priority to a nearby destination on a low frequency, and program other communications depending on the future positions of the terminals.

– In the context of smart buildings, it will also be possible to monitor the health and the exposure of the residents by adjusting the settings of cognitive radio equipment. By tracking the movements of the user, the base stations can readjust and choose their operating frequencies with the aim of limiting the individuals' exposure to radiation. Indeed, a situation where a user is covered simultaneously by several access points on the same frequency is not desirable. However, exposure to radiation on different frequency bands may prove a positive thing. It would enable users to choose the most energy-efficient frequency.

– Exploiting cognitive radio for "smart grids" or smart electrical networks constitutes an interesting perspective for a smart building. Indeed, using cognitive radio, we can interconnect the new generation of electricity counters and then transport the information collected to an energy supplier to optimize consumption in real time. This avenue is being seriously explored at present by electricity providers, and shows great promise to bring about a significant reduction in energy consumption.

The suggestions given here still need to be explored in depth and adapted to the context of each cognitive radio installation for green energy.

5.6. Conclusion

Cognitive radio was initially put forward as a new paradigm for wireless communication, intended to mitigate the lack of radio resources. This is possible because of dynamic, on-the-fly allocation of frequencies. However, thanks to its agility and capacity to intelligently adapt the parameters of communication, cognitive radio enables us to reduce energy consumption and steer wireless networks towards the green ideal. In this chapter, we have suggested a number of avenues to be explored so as to make wireless communications more energy efficient. Our solutions involve using the properties of reconfiguration and adaptation which cognitive radio offers, and cooperation between the different nodes, so as to share information and divide the tasks to be performed.

We believe cognitive radio offers significant opportunities to increase the lifespan of a network and reduce its energy footprint – constraints which, in the domain of ICT, are crucially important. However, in order for such approaches to yield the expected results, it is desirable that there be collaboration and comprehension between the users. Indeed, users have to agree to interaction and cooperation with other users, and sometimes accept lower data rates. Clearly, cognitive radio offers an opportunity, but effort is needed on the part of all the actors in order to seize it.

5.7. Bibliography

[AKY 06] AKYILDIZ I., LEE W.Y., VURAN M.C., MOHANTY S., "Next generation dynamic spectrum access cognitive radio wireless networks: a survey", *Computer Networks*, vol. 50, Issue 13, p. 2127-2159, 2006.

[FCC 02] FCC, Spectrum policy task force report, ET docket, no. 02-155, November 2002.

[GRA 09] GRACE D., CHEN J., JIANG T., MITCHELL P.D., "Using cognitive radio to deliver green communications", *IEEE Crowncom*, Hanover, Germany, August 2009.

[GÜR 11] GÜR G., ALAGÖZ F., "Green wireless communications via cognitive dimension: an overview", *IEEE Network*, vol. 25, Issue 2, p. 50-56, March-April 2011.

[MIC 10] MICHAEL N., MOY C., ACHUTAVARRIER V., PALICOT J., "Area-Power tradeoff for flexible filtering in green radios", *Journal of communications and networks*, vol. 12, Issue 2, p. 158-167, 2010.

[PAL 09] PALICOT J., "Cognitive Radio: An Enabling Technology for the Green Radio Communications Concept", *ACM IWCMC'09*, Leipzig, Germany, 21-24 June 2009.

Websites

[COG] Wun Cognitive Communications initiative: www.wun-cogcom.org.

[EAR] Energy aware radio and network technologies: www.ict-earth.eu.

[GRE a] Greenet: an initial training network on Green Wireless Networks: www.fp7-greenet.eu.

[GRE b] Green Touch initiative: www.greentouch.org.

Chapter 6

Autonomic Green Networks

6.1. Introduction

Autonomic networks are systems which are capable of reconfiguring themselves automatically (self-configuring), constantly seeking to improve their own performances (self-optimizing), detecting, diagnosing and repairing problems with hardware or software (self-healing) and protecting themselves from attacks or cascading failure (self-protecting). Human intervention is limited to guiding the network by providing it with high-level directives [KRI 06]. This paradigm of management can be applied to numerous contexts, from the simplest component to the most sophisticated information technology systems [KRI 08].

Applied to green networks, which aim to reduce their carbon footprint, these new functions would offer them the assurance of functioning efficiently and in an environmentally friendly manner, even under changing conditions.

Chapter written by Francine KRIEF, Maïssa MBAYE and Martin PERES.

6.2. Autonomic networks

The vision of autonomy in networks is the creation of a system which is able to manage itself without human intervention, in order to deal with the increasing complexity and excessive costs of network management today, whilst also laying the groundwork to cater for the needs of the ubiquitous computing of the future [KRI 06]. Thus, networks become a set of self-governing entities which do not require human intervention, except to specify high-level directives and objectives, which enables the details of management and control of the software and hardware components of the autonomic system to be hidden from the administrator. Hence, the concept of autonomic networks is largely inspired by what exists in biology – in particular the autonomic nervous system, to which the discipline partly owes its name [HOR 01]. Indeed, the autonomic nervous system is at the root of a set of activities which the human body performs without conscious knowledge. The nervous system is responsible for regulating the beating of our hearts, the rate of our respiration and many other vital functions. Just like humans with our autonomic nervous systems, a network with its autonomic entities must be reliable and offer guarantees of availability, safety, survival, security and maintenance [STE 03]. To this end, this paradigm aims to bring together and harmonize all research domains which could contribute to the realization of autonomic networks [KRI 06].

The earliest architecture for self-management was put forward by IBM, through the ACI (Autonomic Computing Initiative). In this architecture, the central element is the autonomic element. Today, there are many proposals in circulation for architectures of autonomic networks [AGO 06; AUT; ANA; 4WA] that have emerged since IBM's proposal [JEF 03; JAC 04], but to date, no standardization has been defined.

In an autonomic environment, the components of the network must perform the following self-management functions (also called "self-functions"):

– self-configuring;

– self-healing;

– self-protecting;

– self-optimizing.

These self-functions are the heart of an autonomic element which evolves in its environment. Many other self-functions have now been defined; among the most significant, we can cite self-adaptation, self-organization and also self-awareness, which is knowing the state of the surrounding environment. This capability implies that the autonomic element must undertake phases of production, usage, validation and sharing of its knowledge.

Self-awareness thus adds another level of complexity to the model for realizing autonomy: that of knowledge management. To this end, the "knowledge plane" was put forward, to manage all aspects related to knowledge. It was first proposed by Clark et al. [CLA 03] and has an important part to play in autonomic networks. Indeed, in an autonomic network, the elements must have a certain amount of knowledge of their environment – self-awareness – in order to be able to adapt themselves. This knowledge is either gleaned and constructed by the entity itself or provided by the administrator or its partners.

The knowledge plane provides an infrastructure for the management and exchange of, and reasoning about, the knowledge of the network. It completes and closes the "control loop" by giving the system the capacity to automatically acquire experience and reliability during its activity cycle and to react dynamically to an event that occurs on the network (see Figure 6.1).

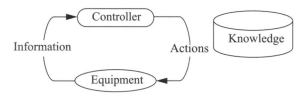

Figure 6.1. *Closed control loop*

In the following sections, we shall describe the four main "self-functions": self-configuring, self-optimizing, self-protecting and self-healing, and highlight the respective advantages to each in the context of green networks.

6.3. Self-configuring

An autonomic network has the capacity to automatically configure itself in order to adapt to changes in the environment or to facilitate other management objectives. Unlike with conventional management – where the function of configuring systems with numerous heterogeneous components is extremely difficult and requires a great deal of expert time – autonomic systems configure themselves in a manner transparent to the operator, following high-level directives and policies. The policies specify the goal which must be attained, but not the way in which the components must be configured in order to attain it. A new autonomic element added into an autonomic system self-configures in a transparent manner, and the rest of the system adapts to its presence, reconfiguring itself if need be. Thus, configuration errors are avoided, and consequently the administrators gain a great deal of time, because they only have to deal with the high-level objectives once the configuration policies have been specified [KRI 06].

6.3.1. *Importance of self-configuring for green networks*

Self-configuring is an important function because it enables an autonomic system to adapt its behavior to its environment. With a network whose goal is to minimize its energy consumption, such as a wireless sensor network, this function allows it, once it is installed or when new sensors are added, to self-organize so as to ensure a data path which minimizes the energy consumption. A distributed clustering algorithm based on numerical classification would be particularly well-adapted to this context [KRI 08].

The function of self-configuring is closely linked to the other self-functions. In order to continue to minimize its energy consumption and thus optimize its mode of functioning, an autonomic green network must be able to reconfigure itself many times without causing a service interruption [HOS 11; MBA 11]. When a fault occurs, in order to heal itself, the network will also have to reconfigure one or more autonomic elements so as to be able to continue to guarantee function, in either normal or downgraded mode [YOO 09]. Finally, in order to protect itself, the autonomic green network must also be capable of reconfiguring itself dynamically [PER 11].

6.4. Self-optimizing

Self-optimizing is one of the essential functions of autonomic systems [STE 03]. It consists of the activity and capacity of a system which attempts to automatically maximize the use of available resources, and its performances, in light of clearly-defined metrics, representing the criteria of performance. This is not simply a question of choosing the correct configuration parameters for a system, but also adjusting its internal functioning. This function poses a three-pronged problem, relating to:

– business constraints which are the high-level objectives of the network designer or the administrator. The system should not make choices or perform actions which run counter to these constraints. As examples of constraints, one might cite the contracts entered into with customers (Service Level Agreements), or those related to the type of network (in a surveillance network, the objective is to detect any intrusions);

– metrics: these are the parameters for measuring the quality of resource usage while conforming to the constraints. There are usually a number of metrics, which may be mutually interdependent. For instance, the classic metrics of QoS, which are bandwidth, delay, jitter and loss ratio, are interdependent or contradictory. Decreasing the loss ratio by increasing buffers of intermediary nodes in the core of a network may lead to an increase (degradation) in terms of delay, for instance. In such cases, it is crucial that the constraints define – either implicitly or explicitly – the order of metric preference or priority for optimization;

– and, finally, available resources. The easiest way to satisfy the constraints is to over-dimension the system by giving it more resources than are actually needed for the intended activity. However, from a commercial point of view, this approach is less beneficial. The usage ratio of the resources, in combination with the metrics, should enable us to construct an order relation for all states of the system, so that the system has an objective barometer to measure or evaluate its own state. Thus, the system will improve and fine-tune itself as and when required, based on the result of its self-evaluation on the readjustments which it carries out.

Figure 6.2 illustrates the relationships between all the concepts developed above.

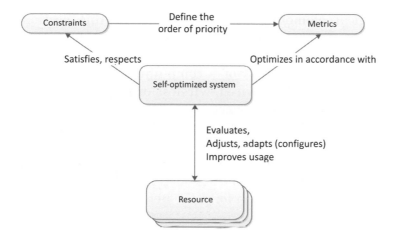

Figure 6.2. *Elements of a self-organized system*

Self-optimizing can also be divided into self-adjusting and self-tuning functions [SAL 09].

6.4.1. *Self-optimizing for green networks*

Research on self-optimizing, with a view to green operation, can be divided into three categories, which we shall discuss below:

– self-optimizing of network architectures for green usage;

– self-optimizing of communication protocols and paradigm for green usage; and

– self-optimizing of QoS mechanisms for green usage.

6.4.1.1. *Self-optimizing of green network architectures*

Network architecture is a generic term which denotes a set of functional elements describing the organization, interaction and functioning of the components within a system for establishing communications. In this section, when we use the term network architecture, we refer to the topological elements. In other words, in this section, we consider an architectural element to relate to the organization of the network equipment (arrangement of the nodes, organization of communications, etc.), the technological choices (wired, wireless or mobile network, etc.) which can have an impact on the energy balance of the systems. Research projects about optimized architectures to minimize energy consumption have developed primarily along three main axes: wireless local area networks (WLAN), mobile networks (GSM and its successors) and architectures of the data centers which form the heart of the Cloud.

The research community has devoted a great deal of attention to the question of energy management in the architectures of wireless networks [TSE 11; AMI 09; QIA 05; BAI 06]. Such research has given rise to many energy-efficient technologies such as Bluetooth, IEEE 802.15.4. However, the initial objective of these technologies was to prolong energy autonomy of terminals on which they were installed, rather than necessarily trying to reduce the carbon footprint. In this scenario, energy is an inherent constraint for availability of service, whereas in the green approach, it can be an additional constraint, imposed in order to reduce the system's carbon impact. Jardosh *et al.* [JAR 09] show that in most major WLANs, a large proportion of the access points (APs) remain on standby without being used, and propose a solution of availability of resource on demand. Agrawal *et al.* [AGR 10] construct a model which illustrates the energy consumption of the stations (STAs) in Continuously Active Mode (CAM) or static Power Save Mode (PSM) for TCP traffic. These studies show that in architectural terms, the deployment and insistence on a near-permanent availability of service lead to a considerable wastage of energy. Self-optimizing of these architectures could consist of putting in place a decentralized system which, depending on the demand on the network and the needs of the users, autonomously adjusts which APs are switched on or off, while maintaining coverage for all the users at all times. The architectures of wireless sensor networks are an example of wireless architecture which has arisen from the energy constraint. Many works in this domain deal directly with self-optimizing of the activities on the network with a view to saving energy in routing [KAS 09; WEN 08] for example.

With the densities of mobile phones and ever more readily-available smartphones, and the resultant mass installation of BTSs in large urban areas, the consumption of mobile devices represents nearly 25 kg of carbon per year per device [ERI 07; EZR 09]. This consumption is largely due to the repeated scanning as devices search for a network [ONG 11]. A number of approaches to remedy this problem have been put forward. Ong et al. [ONG 11] propose an architecture to cater for the anticipated convergence towards IP, which is based on the broadcast of contextual information by the base transceiver station so as to reduce the terminals' need to scan. Hossain et al. [HOS 11], on the other hand, propose a cooperative, self-optimizing architecture based on cooperative and intelligent decision-making by the BTSs so as to switch from one power mode to another depending on the traffic conditions on the networks. Ezri et al. [EZR 09] propose the idea of green cells with dedicated receiver base stations, so as to minimize emissions from the mobile stations. Also, the base station [SAK 10] can readjust based on peak periods in sleep mode and its periods in activity mode.

The carbon impact of network services stems largely from the consumption of servers in the data centers and the cooling systems which accompany them since the advent of the Cloud. One solution to this problem emerged a number of years ago: the idea is to encourage sharing of resources with virtualization. A large proportion of the work done on green energy is aimed at finding a solution to the energy consumption of the servers in data centers. Firstly, there is the energy consumption because of memory access; thus, Khargharia *et al.* [KHA 09] showed that by using a system to selectively put the hard disks on standby, a 25% gain in energy could be achieved. Then, there is the architecture of Cloud services [BAH 10], which poses a genuine problem because of the conflicting objectives: high availability (guarantee) and energy efficiency (optimization). Zheng *et al.* [ZHE 11] propose a solution based on an architecture which optimizes the usage of data centers by a system for distributing the workload. This distribution of the workload involves routing requests to the nearest services.

In all these approaches, self-optimized architectures – i.e. architectures which are capable of adjusting their energy-consuming activities in order to minimize them – seems to be the direction in which all the solutions are converging for future green networks.

6.4.1.2. *Self-optimizing of protocols and the paradigm of green communication*

The execution of network protocols also has a not-insignificant impact on energy expenditure. Indeed, Christensen *et al.* [CHR 05] show that certain protocols are unnecessarily very communication-intensive. ARP requests, for instance, may account for up to 50% of the packets in an LAN connection, in spite of the existence of caches. In the core of the network, transfer is responsible for nearly all energy consumption, including 66% for IP routing and 11% for the commuted network [LYO 08; BO 11]. Works aimed at minimizing the carbon impact of data transfer protocols have begun to emerge. Cianfrani *et al.* [CIA 10], for instance, propose an extension of the OSPF protocol which enables certain links to be switched off in periods when there is low traffic on the network. They demonstrate that according to the topology, up to 60% of links could be shut down without negatively impacting the performance of the network. The works of Chu *et al.* [CHU 11] in the same vein show that by equipping the GMPLS protocol with functionalities enabling it to place certain routers in sleep mode, it is possible to reduce energy consumption by up to 15% without affecting the performances.

All these initiatives show that new-generation protocols must integrate a process of self-optimizing, while still respecting their conventional constraints and factoring the energy consumption into their performance metrics. Thus, they must be able to fulfill their primary functions whilst minimizing the overall carbon impact of the system, in a decentralized manner on the different components of the system.

6.4.1.3. *Self-optimizing of QoS mechanisms for green networks*

In order to be able to respect the contracts signed with their customers, network access providers often resort to over-dimensioning and duplication of the infrastructure so as to ensure a high degree of availability. QoS mechanisms should be able to progressively eliminate these practices, which almost double the energy consumption of the network resources. Allocation of resources, which is at the root of most QoS mechanisms, also tends to under-use resources, which leads to a waste of time and energy.

Research projects devoted to green QoS mechanisms are still thin on the ground. However, one of the most significant advances is the extension of the standard IEEE 802.11e for energy management: APSD (Automatic Power Save Delivery) [PÉR 10]. APSD uses the QoS mechanisms contained in the standard IEEE 802.11e to reduce the signaling load. This mechanism functions in two modes: unscheduled mode (unscheduled APSD (U-APSD)) and scheduled mode (scheduled APSD (S-APSD)). Another approach to optimization of the QoS mechanisms [SCH 11] consists of allocating resources on the users' demands.

Self-optimizing of the mechanisms with a view to economizing energy will be one of the most important issues in the domain of QoS but also one of the greatest challenges to overcome because of the compromise which it requires, as Liu *et al.* [LIU 11] point out.

6.5. Self-protecting

Self-protecting is a self-function whose aim is to render an autonomic system more resistant to its environment. Many kinds of environmental disturbances may prevent an autonomic system from working properly. We can classify these disturbances depending on the layer or layers of the OSI model which they affect, but also whether or not they are deliberate (malicious). An autonomic system is said to be resistant to involuntary disturbances if it has a good chance of functioning in its normal environment. We say that an autonomic system is resistant to deliberate disturbances if it continues to function (whether in degraded mode or otherwise) in a deliberately hostile environment.

Here are some examples of disturbances which a wireless autonomic system may find itself faced with:

– PHY/MAC layer: electromagnetic/cosmic ray disturbances, multipath (constant or dynamic), collisions;

– network layer: duplicate IP addresses, incorrect routing, selective routing (Byzantine nodes);

– application layers: corruption of application data such as the data of a wireless sensor network.

These disturbances cause nuisances which prevent the autonomic system from functioning optimally. These nuisances have either a direct or indirect impact on the latency of the communications but also on the energy consumption of the system. By identifying the vulnerable points present in each layer and proposing an architecture which takes account, as far as possible, of these problems (depending on the need), it is possible to increase the self-protecting of an autonomic system.

The solutions proposed can be classified into two categories:

1) local solutions: a solution is said to be local if there is no need to disturb the environment in order to resolve the problem. A local solution therefore should not require communication with the outside world; there are two sub-categories of such solutions:

- passive solutions: a local solution is said to be passive if it does not add to the autonomic system any particular treatment to deal with the disturbance (e.g. tropicalization varnish),

- active solutions: a local solution is said to be active if it requires an additional treatment in order to detect/repair an error due to a disturbance (e.g. error-correcting code);

2) global solutions: a solution is said to be global if a protocol has to be established between several nodes (e.g. collaboration) to mitigate/resolve the problem caused by the disturbance.

In the rest of this section, we shall examine disturbances which impact the executive support system. This will be followed by an analysis of the constraints imposed by energy sources. Then, we shall look at disturbances relating to inter-node communications. Finally, we shall explore the case of deliberate disturbances aimed at corrupting an autonomic system.

6.5.1. *Protection of the executive support*

The hardware on which autonomic systems run is sensitive to disturbances. If a system is able to effectively protect itself against these disturbances, it can increase its lifespan and therefore offset the pollution generated in its production over a longer lifecycle. In an extreme case, the system could not work at all, which would cause its production cost/utility ratio to tend toward infinity.

The immediate environment disturbs the proper function of the hardware supporting an autonomic system. Its impact must therefore be understood and taken into account when defining the objectives of the system and its self-protecting mechanisms. The disturbances may be chemical, electromagnetic or ionizing in nature.

6.5.1.1. *Oxidation and corrosion*

Any naked electronic circuit tends to oxidize with time, which decreases the quality of the electrical contacts and alters the behavior of the circuit. This is even more critical in wireless communications where resistances are minutely calculated to tune the antennas to one another, in order to increase the signal-to-noise ratio and therefore decrease packet losses. This phenomenon of oxidation is even more present outdoors, where the temperature, humidity and salinity of the air are not controlled. In order to guard against these phenomena, it is possible to isolate the circuit in a hermetically-sealed casing. If the electronic circuit contains an antenna, it is important for this box not to be conductive, so as to avoid the Faraday cage effect, which would prevent the antenna from transmitting out of the box but would also prevent it from receiving communications originating elsewhere. It is, of course, possible to leave the antenna outside of the box, in which case its conductance would have no effect on the radio. In conditions of extreme humidity, it is also possible to treat the circuit with a tropicalization varnish [ELE]. This varnish renders the circuit completely impermeable, and protects it even in case of immersion in water. In addition to protecting against corrosion and oxidation, tropicalization varnishes prevent the phenomenon of "metal whiskering" [WHI]. This phenomenon, which is not well understood, is to blame for the appearance of micro-whiskers growing vertically on the surface of metals such as zinc or tin. This phenomenon mainly manifests itself in the case of physical stress on the metal, whether due to major temperature differences or to any other deformation of the surface. However, the application of a tropicalization varnish requires heating the circuit, for between four to eight hours, to a temperature of around 90°C to get rid of all traces of humidity [ELE]. Because this stage is costly in terms of energy and increases the carbon balance as well as the final price of the system, it should only be applied if absolutely necessary. These solutions are local and passive.

6.5.1.2. *Electromagnetic radiation*

Electromagnetic radiation is increasingly present in our society. It is generally the result of a current passing through a wire which, in reaction, begins to give off radiation like an emitting antenna. This radiation is thus generated either deliberately or involuntarily, as is the case in electrical motors. When this radiation comes into contact with an electrical conductor, it generates voltage within it. This is the principle behind a receiving antenna. This voltage is added to that which is already present, and "noises" the voltage obtained. If the conductor is a power supply cable, too great a variation in the voltage can cause errors in the computations taking place on the autonomic system. Conversely, if the conductor is a data line, the bit being transmitted at that moment may be altered from 1 to 0 or vice versa. However, this type of radiation can be stopped by surrounding either the conductor or the whole autonomic system with a conductor which is wired to the ground of the system. We then say that this conductor or system is shielded, by the Faraday cage principle. In the case of wired data transmissions, it is also possible to reduce the transmission speed so as to decrease the probability of error. Thus, it is relatively easy to guard against this type of radiation. However, such protection increases the cost of the system because more primary materials are needed, which has a financial cost and an "equivalent carbon" price. The solution to counter electromagnetic radiation is local and passive.

6.5.1.3. *Ionizing radiation*

Besides electromagnetic radiation, circuits are also influenced by ionizing radiation. This radiation may stem from human activities – particularly nuclear activities – but can also come from the cosmos. On the ground, we are relatively well protected from cosmic radiation by the ionosphere, which lies at an altitude of over 60 km; however, the same is not true for systems such as the international space station, which orbits at an altitude between 330 and 410 km. This radiation is often overlooked by computer scientists, although it can explain a number of software problems. Surprisingly, relatively few studies on it exist, even though our society is increasingly dependent upon computers in order to function. This problem has been lampooned on the Website *Le Point* [LEP], which cites the case of a Microsoft study, or of bugs in electronic voting machines. The aim of the "Altitude SEE Test European Platform" (ASTEP) project is to characterize nuisances and evaluate the probability of these disturbances occurring. This is a joint academic and industrial project, financed by STMicroelectronics, JB R&D and the L2MP-CNRS. Its results show that the effects of this radiation are primarily "bit-flips" (changes in the state of a bit) with an equal probability of $0 \rightarrow 1$ or $1 \rightarrow 0$ transitions. It has also been shown that the altitude and the etching width exert an influence on the probability of occurrence of these bit-flips.

At an altitude of 2,552 m and with 9,000 hours of exposure, 60 events have caused 90 bit-flips on 5 billion memory cells. The most conventional solution to counter this problem is to use memories with error-correcting codes (ECCs). However, these memories are more expensive, and are generally used only for servers. For high-altitude onboard autonomic systems, which are generally of crucial importance, it is not enough to simply protect the memory: we must also protect the data buses and the records of work, because the probability of their being affected increases with altitude. One solution, in order to avoid having to design a fault-resistant processor, is to have redundancy in the computation systems, and then use a majority vote system [SKL 76]. These systems have been studied, and are used in high-altitude missions such as for American space shuttles. The disadvantage to adding redundancy in the hardware is the increased manufacturing cost as well as the increased energy consumption when in use. Hence, these solutions should only be adopted for critically important systems; other systems could perhaps make do with software solutions to detect inconsistent data. The possible solutions to protect against this radiation are local and active.

6.5.1.4. *Conclusion*

An autonomic system cannot protect itself from its environment if it is unaware of its executive support and the constraints which that support entails. In the same way that network protocols are designed to withstand the loss of network packets, the software must be aware of the limitations of its executive support, and must put an architecture in place to reduce the risk of corruption of its state. Consequently, we have to know the execution environment of the autonomic system in order to choose the appropriate hardware and the software architecture for the system.

6.5.2. *Protection of the energy source*

Autonomic systems are subject to significant constraints in terms of energy supply, because they are supposed to function by themselves, without human intervention. This dependency has to be taken into consideration at the design stage, so that the system can adapt to the energy resources available to it, but also to its objectives. Each of a system's actions requires energy – usually electricity. The energy is then transformed in order to carry out the action. Given that this transformation is performed with a certain yield, parasitic energy is created. For electrical systems, we generally obtain heat by way of the Joule effect and emissions of photons (generally infrared) by way of the black body effect. Hence, we can say that in order for an autonomic system to consume little, it must do as little as possible.

6.5.2.1. *Management of consumption and temperature*

Since a system consumes more energy as it works, we can say that the faster a system works, the hotter it will get. If that heat injection exceeds the system's capacity to cool itself (heat budget), its temperature will increase and decrease its lifetime. If that temperature increase becomes too great, it can lead to the self-destruction of the system. It is also increasingly common for a system to have to indicate its power budget (maximum consumption). Once the power budget has been determined and respected, the system can guarantee a certain lifetime when it is fed by an energy source with fixed capacity, such as a battery. For these reasons, it is sometimes necessary for a system to internally model the energy costs associated with every possible action so as not to exceed its energy and heat budget. In the case of a node in a wireless sensor network, we can show several heavily consuming components as well as several states in which they may find themselves. For instance, the consumption of the radio will be different depending on whether it is receiving, transmitting or in sleep mode. By prioritizing certain actions, it is possible for a system to remain within its energy budget, at the cost of an increase in latency before certain operations. Hence, the solution is local and active.

6.5.2.2. *Loss of power supply*

Precisely because of their autonomy, autonomic systems are often forgotten by users. When power supply is lost, a system is no longer able to carry out its functions, which can cause serious malfunctions in systems connected to it, more or less in the long term. Hence, the administrator must be alerted before a failure can have an impact on the functions of other systems. This constraint of state signaling necessitates that a system advises the administrator when a break in the electrical current occurs. Indeed, whatever the power source used by an autonomic system, it will always experience some periods of downtime, of greater or lesser duration. Such cutoffs may result from a failure on the electrical grid, a faulty battery or indeed an act of sabotage intended to render the system non-operational (denial of service). It is possible to guard against these losses of power supply by storing electrical energy internally. There are two usual storage solutions:

– condensers: low density (~ 5 Wh/L), low capacity, highly cyclical and relatively strong charge/discharge current;

– batteries: high density (~ 300 Wh/L), high capacity, not highly cyclical and a relatively weak charge/discharge current.

Hence, condensers are relatively well adapted to deal with micro-cutoffs, but their low energy density means they cannot be used to cope with prolonged supply interruptions. Batteries, on the other hand, are limited to long-term energy storage because of their feeble discharge current. Their high density and energy capacity and their low cost account for their success.

Because all the components of an autonomic system must be able to alert the administrator in case of a serious or prolonged malfunction, certain choices must be made concerning the architecture of the supply system. In case of a supply failure, the autonomic system must therefore be able to perform the following operations:

– saving its state in a non-volatile memory, so that it can resume its processing;

– alerting the administrator and any neighboring systems of its breakdown. Thereby the administrator is alerted earlier, and this enables the neighboring systems to update their routing tables;

– putting itself in sleep mode and waking itself up periodically until a new energy source has recharged its internal source.

Given that these operations are costly in terms of both time and energy, each autonomic system must have a sufficient onboard power source to power itself not only throughout the duration of those operations but also throughout the standby period. Hence, it is necessary to specify a maximum lifetime and the consumption in this downgraded mode in order to calculate the energy necessary as well as the technological solution capable of delivering it.

6.5.2.3. *Energy-aware network layers*

In a point-to-point communication, the overall energy consumption of an autonomic network can be broken down into two parts:

– transmission: energy cost of communication on the transmitter node;

– reception: energy cost of communication on the receptor node(s).

Because the energy consumption for reception of a message is different from that for transmission [MAR 09], the consumption is not distributed evenly between the different networking nodes. In the case of a wireless network, the sending of a message forces all the nodes within range of the transmitter to consume energy to at least partially decode the message so as to identify whether or not they are the intended addressee. This non-homogeneous distribution of energy consumption poses problems in terms of the lifespan of the network. In order to lengthen the lifetime of the whole network, the energy consumption must be as low as possible but should also be distributed homogeneously. To reduce its consumption and disturbances of the nodes that are not involved, it is in the interests of a transmitter to decrease its transmission power [GOM 07]. However, if the nodes are not within radio range, the information has to be routed by way of intermediary nodes. The selection of the number of hops to perform to connect nodes which wish to communicate has been the object of many studies, such as those done by Sikora et al. [SIK 04], and has even been modeled by Wang et al. [WAN 06]. The results of these studies show that short-range communications decrease the number of retransmissions necessary (increasing the signal-to-noise ratio) and reduce disturbances on nodes which are not involved. In addition, as the transmission time increases in a linear fashion with the number of intermediary nodes, problems relating to the Quality of Service (QoS) begin to manifest themselves. Routing algorithms which take these issues into account have been put forward – e.g. those advanced in Akkaya et al. [AKK 03]. This algorithm uses a model of the energy cost and the latency of transmissions to optimize the consumption and end-to-end delay. This algorithm also takes account of the level of charge of the battery, so as to distribute the consumption correctly.

6.5.2.4. *Putting part of the network on standby*

Distributed autonomic systems with a high degree of redundancy do not need all the nodes to be active at once in order to function normally. Therefore, it is possible for certain nodes to enter sleep mode in order to decrease the average energy consumption of the network. These nodes must not be deactivated to the detriment of the QoS [ZHA 06] or the resistance of the network [WAN 08].

6.5.2.5. *Conclusion*

Any autonomic system must be aware of the energy involved in its actions so as not sabotage its own operational objectives (lifespan), heat specifications (energy/heat budget) or QoS objectives. Hence the system must contain a model of itself which allows it to adapt to stresses so as to protect its resources. Because of the presence of this model, the solutions put forward to deal with the energy problem are active, local for the physical part and global in the case of the network layers.

6.5.3. *Protection of communications*

An autonomic system acquires and processes information. However, the data created or received are generally not used on site by the users. Therefore, those data must be channeled to the users. It is not enough simply to protect the autonomic system's hardware, because in order to be useful, the data must reach the users without suffering alterations and with as little latency as possible.

6.5.3.1. *The resistance of the hardware / MAC layer*

6.5.3.1.1. Arbitrary collisions and disturbances

In the context of networks, using a shared transmission medium, as is the case with wireless networks, when two nodes communicate at the same time on the same frequencies, they jam the nodes within range of the two transmissions. Thus, these nodes can receive neither one nor the other of the communications. Many possible solutions to these problems have been put forward. Historically, CSMA/CD was introduced in order to reduce the probability of collisions in wired networks. However, this solution does not apply in wireless networks because not all the nodes are necessarily within range of each other. The CSMA/CA protocol proposes to put in place a signaling system to enable the nodes to synchronize with one another in order to avoid collisions. An extension was advanced for multi-hop *ad hoc* networks [CHO 10] because CSMA/CA causes needless collisions when multiple hops separate the nodes which need to communicate.

6.5.3.1.2. Multipath

In communication between two radio nodes, the path followed physically by the waves may be direct, diffracted or refracted. In the absence of external disturbances, the signal will be received multiple times by the receptor node, with delays dependent upon the length of the paths taken and powers dependent on the distance and on the number of reflections experienced by the signal. This phenomenon is called multipath. We say that it is static or dynamic depending on the mobility of the environment. In a static environment, these delays and levels of signal absorption only vary slightly. With a dynamic environment, as is the case in an urban area, the paths followed by the signal may partly change. This phenomenon causes a decrease in the signal-to-noise ratio (SNR) and therefore poorer quality reception. There are many solutions to improve the signal-to-noise ratio. These solutions may be local, as is the case when channel equalizers are used [TON 91] or global, requiring a different type of encoding, as is the case with the orthogonal frequency-division multiple access [MOR 07] or needing multiple antennas, such as for 802.11n (MIMO WiFi)

[SHA 06], which takes advantage of multipath to increase data rates.

6.5.3.2. *Protection against internal and external attacks*

When autonomic systems handle data with value for certain individuals, the temptation may be great for these people to monopolize or falsify them. By so doing, they endanger the correct functioning of the system.

6.5.3.2.1. Human attack on the medium – malicious routing

Self-configuring lightens the workload of an autonomic network administrator. However, when it comes to routing, attackers can take advantage of this automation to divert all of the network's communications to a computer which they control. In the context of an autonomic network for intrusion detection, it is possible to extend this attack so as to be able to filter out only the alarm messages being sent to the administrator, rendering the network completely useless. Routing security is discussed in detail in Karlof *et al.* [KAR 03].

6.5.3.2.2. Authentication and confidentiality

When a message is received, a network node determines the source of the packet by reading the source IP address. However, that address may have been modified during routing, or forged by the transmitting node so as to appear to be a different node. For these reasons, authentication of the source of packets cannot be done based only on the IP address or MAC. Most authentication methods are based on the use of local secrets at the nodes and on symmetrical or asymmetrical cryptography. For autonomic systems for which energy is not a real constraint, the current trend is toward the use of Transport Layer Security [IEF]. TLS provides both confidentiality and authenticity. For systems for which it is more of an issue, asymmetrical cryptography has long been considered too greedy in terms of energy resources and of live memory [PER 02]. However, new research has shown that in certain cases, asymmetrical cryptography is possible, particularly thanks to the use of elliptic curve cryptography [BLA 05]. A study into the energy impact of cryptography was conducted by Wander *et al.* [WAN 05].

6.5.3.2.3. Corrupted application data

Whether it be by injection, routing or simply by corruption of the nodes of an autonomic network, an attacker can influence the data read by the system and manipulate them to his own advantage. If an attacker has found a fault in the security system, the final rampart to detect the modifications is the post-processing of the data outside of the system [MOY 09]. It is also possible to use trust and reputation to detect malicious/corrupted nodes [MOY 09].

6.5.3.3. *Conclusion*

Because communication is of absolutely crucial importance for autonomic systems, guaranteeing the security properties and a minimum Quality of Service increases the trustworthiness of the system (in the discipline, one speaks of "trustworthy computing"). The solutions proposed above are all active and global because they require collaboration between the nodes to share the medium.

6.6. Self-healing

The function of self-healing consists of detecting malfunctions, diagnosing and carrying out appropriate actions without interrupting the functioning of the system. Self-healing elements and applications have to be capable of observing system failures, evaluating the stresses of the outside environment and applying the appropriate corrective measures. In order to automatically discover system malfunctions or possible failures, we have to be aware of the expected behavior of the system. An autonomic system has a knowledge base which should enable it to determine whether the current behavior is consistent and predictable in the context of the environment. In new contexts or in different scenarios, new system behaviors may be observed, and the knowledge module must be able to evolve with the changing environment.

In order to be able to detect and then correct possible malfunctions of the system, the function of self-healing follows a process which is illustrated in Figure 6.3.

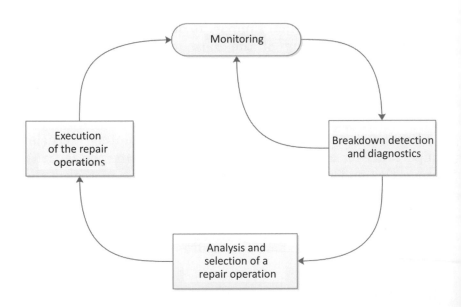

Figure 6.3. *The process of self-healing*

The first stage in the process is called "monitoring". During this stage, the task of the autonomic entity is to detect any abnormal behavior. Once the inspection is completed, the observation data collected are sent to the next phase. The second stage is called "breakdown detection and diagnostics". If the diagnostic report indicates that no malfunction has occurred, we go back to the previous phase. If there is a fault detected, the errors are analyzed in the next stage, and a repair or "healing" operation is chosen. Any healing necessary is carried out in the next stage. Once the faults have been repaired, the process of self-healing can begin again. This process constitutes a closed control loop [SHA 07].

While numerous research projects about autonomic networks have been realized, few of these deal specifically with self-healing. One might cite the works of Lu *et al.* [LU 11], who propose a self-diagnosis function able to identify the origin of breakdowns in a telecommunications network.

In the following sections, we illustrate the function of self-healing for two particular types of network: wireless sensor networks, which by nature are constrained in terms of energy, and smart grids.

6.6.1. *Application to wireless sensor networks*

A wireless sensor network may be considered a particular *ad hoc* network, generally made up of a very high number of nodes, called sensors, which present significant restrictions in terms of resources (memory, computation capacity, energy). This type of network is usually devoted to a particular application (monitoring of an area, etc.). The nodes in the network have the primary task of collecting, processing and disseminating information. To do so, they generally have several types of sensors (seismic, thermal, visual, infrared, audio, etc.) which allow them to monitor their environment (temperature, rate of humidity, movement of vehicles, luminosity, atmospheric pressure, noise levels, presence or absence of objects) or physical characteristics such as velocity, the direction of motions and the size. A large number of domains of application are concerned (the military domain, the environment, health, trade and industry, home networks, etc.).

6.6.1.1. *The importance of energy in wireless sensor networks*

The battery is a very important component in a sensor. Usually it is neither replaceable nor rechargeable. It can be powered in part by a power-generating unit, such as solar cells. It is small, and therefore provides a very limited amount of energy. Therefore, it limits the lifespan of the sensor and influences the overall function of the network, which loses nodes over time. For this reason, energy-saving protocols have been an important area of research in the domain of wireless sensor networks.

A sensor uses its power to perform three main actions: acquisition, communication and data processing. The energy consumed for data acquisition is not enormous. Nevertheless, it varies depending on the phenomenon being observed, and on the type of monitoring being carried out. The energy consumed in computation is far less than the energy used for communication. Indeed, it is the task of communication which is the most energy-hungry. This encapsulates communications in both transmission and reception modes. A good energy management scheme therefore has to prioritize the efficiency of communications.

6.6.1.2. *The importance of self-healing in wireless sensor networks*

When we consider large wireless sensor networks, which usually contain hundreds or even thousands of nodes, these are essentially considered to be "disposable" and redundant, meaning that if one of them dies or fails, it need not systematically be replaced. In addition, these nodes are often deployed in hostile environments where it is difficult to intervene. We suppose therefore that the nodes have a finite power reserve to begin with, and that they die when that power reserve is exhausted. The protocols and the application software must, therefore, be developed to include fault tolerance mechanisms. In this context, the implementation of a self-healing function is essential in order to ensure the correct function of the network, which otherwise would risk being utterly paralyzed when a node is no longer functional.

6.6.1.3. *Example of self-healing function*

In [YOO 09], the authors put forward a model of faults and a self-healing architecture for a ubiquitous wireless sensor network. Resumption of service after an incident can be ensured by restart/reprogramming of the defective sensors, which extends the lifespan of the wireless sensor network. This approach is original because most published research deals with this problem by isolating the defective nodes.

The basic fault model, associated with a service, identifies four categories of breakdowns: faults in service (QoS), software errors, hardware breakdowns (relating to the battery or the machine's hardware), and faults related to the environment.

The self-healing architecture proposed considers different types of nodes: the sensors and the sink, which gathers the observation data. The sink is supposed to have more computation and communication resources at its disposal. The self-healing function situated at the nodes is charged only with monitoring the state of the system and executing the adaptation strategy. For this purpose, the sensor has a diagram of state which describes its normal function. When an error is detected, the information gathered is sent to the sink, which generates an adaptation strategy and a repair code destined for the defective node [YOO 09].

6.6.2. *Application to smart grids*

Intelligent electrical distribution networks, known as smart grids, arise from the union between public electrical grids and new information and communication technology with the aim of optimizing production, distribution, consumption and relations between producers and consumers of electricity [MOS 10]. Smart grids help save energy, increase the security of the network and reduce costs. Thus, they contribute to decreasing greenhouse gas emissions.

6.6.2.1. *The importance of self-healing in smart grids*

Numerous studies have been carried out on the subject of smart grids – particularly in the USA – and one of the most important functions of these networks is their self-healing capacity. Duke Energy, one of the largest electricity companies in the United States, uploaded a video on its Website, highlighting the self-healing function with which smart grids are equipped [DUK 09]. Sensors are placed on certain electrical lines so that they can detect any malfunction. When an electric cable is damaged, the self-healing function has to reroute other sources of electricity so as to minimize the impact on the area in question, until repair teams – which it has also alerted – are able to intervene.

6.7. Conclusion

Numerous research works have dealt with the question of autonomic networks. Today, solutions exist for certain types of networks and certain categories of problems. Yet the demand is still high, because networks are constantly becoming more complex and continue to invade our daily lives.

A great many contemporary research projects are investigating solutions for green networking. Such research projects are rooted both in the world of academia as well as in industry. The reasons are many: an increase in the cost of energy, global warming, the image of the company, and so on.

These two approaches naturally go hand-in-hand, given that taking account of the energy constraint only increases the complexity of network management. There again, a certain number of solutions have been put forward, but to date, few have been realized. However – particularly in view of the economic pressure, and with the anticipated increase in wireless sensor networks, which are by nature low-energy – we can expect to see operational solutions emerge in the near future.

6.8. Bibliography

[AGR 10] AGRAWAL P., KUMAR A., KURI J., PANDA M.K., NAVDA V., RAMJEE R., PADMANABHAN V., "Analytical models for energy consumption in infrastructure WLAN STAs carrying TCP traffic", *Proceedings of the International Conference on Communication Systems and Networks (COMSNETS)*, January 2010.

[AKK 03] AKKAYA K., YOUNIS M., "An energy-aware QoS routing protocol for wireless sensor networks", *Proceedings of the 23rd International Conference on Distributed Computing Systems Workshops*, p. 710-715, 19-22 May 2003.

[BAH 10] BAHSOON R., "A framework for dynamic self-optimization of power and dependability requirements in green cloud architectures", *Proceedings of the 4th European Conference on Software Architecture (ECSA'10)*, ALI BABAR M. and GORTON I. (eds.), 510-514, Springer-Verlag, Berlin, Heidelberg, 2010.

[BAI 06] BAIAMONTE V., CHIASSERINI C.F., "Saving energy during channel contention in 802.11 WLANs", *Proceedings of Mob. Netw. Appl.*, 11(2):287-296, April 2006.

[BLA 05] BLAß E., ZITTERBART M. "Towards acceptable public-key encryption in sensor networks", *Proceedings of the 2nd International Workshop on Ubiquitous Computing (ACM SIGMIS)*, p. 88-93, 2005.

[BO 11] BO Y., BIN-QIANG W., ZHI-GANG S., YI D., "A green parallel forwarding and switching architecture for green network", *IEEE/ACM International Conference on Green Computing and Communications (Greencom'11)*, p. 85-90, 2011.

[CHO 10] CHOI J.I., JAIN M., KAZANDJIEVA M.A., LEVIS P., "Granting silence to avoid wireless collisions", *Proceedings of the 18th IEEE International Conference on Network Protocols (ICNP)*, 2010.

[CHR 05] CHRISTENSEN K., GUNARATNE C., NORDMAN B., "Managing energy consumption costs in desktop PCs and LAN switches with proxying, split TCP connections, and scaling of link speed", *International Journal of Network Management*, vol. 15, no. 5, p. 297-310, September/October 2005.

[CHU 11] CHU H.W., CHEUNG C.C., HO K.H., WANG N., "Green MPLS traffic engineering", *Australasian Telecommunication Networks and Applications Conference (ATNAC)*, 2011.

[CIA 10] CIANFRANI A., ERAMO V., LISTANTI M., MARAZZA M., VITTORINI E., "An energy saving routing algorithm for a green OSPF protocol", *INFOCOM IEEE Conference on Computer Communications Workshops*, 2010.

[CLA 03] CLARK D., PARTRIDGE C., RAMMING J.C., WROCLAWSKI J.T., "A knowledge plane for the Internet", *SIGCOMM03*, August 2003.

[DUK 09] DUKE ENERGY, Smart Grid: Self-Healing network: http://www.youtube.com/watch?v=3BF02P9jrKU&NR=1, 2009.

[ERI 07] ERICSSON, Sustainable energy use in mobile communications, White Paper, EAB-07:021801 Uen Rev B:1-23, 2007.

[EZR 09] EZRI D., SHILO S., "Green Cellular. Optimizing the cellular network for minimal emission from mobile stations", *IEEE International Conference on Microwaves, Communications, Antennas and Electronics Systems, COMCAS 2009*, 2009.

[GOM 07] GOMEZ J., CAMPBELL A.T., "Variable-range transmission power control in wireless ad hoc networks", *Mobile Computing, IEEE Transactions*, vol. 6, no. 1, p. 87-99, January 2007.

[HOR 01] HORN P., "Autonomic computing: IBM perspective on the state of information technology", IBM T.J. Watson Labs, New York, *presented at AGENDA 2001*, Scottsdale, October 2001.

[HOS 11] HOSSAIN M.F., MUNASINGHE K.S., JAMALIPOUR A., "An eco-inspired energy efficient access network architecture for next generation cellular systems", *Proceedings of WCNC'2011*, p. 992-997, 2011.

[JAR 09] JARDOSH A.P., PAPAGIANNAKI K., BELDING E.M., ALMEROTH K.C., IANNACCONE G., VINNAKOTA B., "Green WLANs: on-demand WLAN infrastructures", *Proceedings of Mob. Netw. Appl. 14*, 6 December 2009.

[KAR 03] KARLOF C., WAGNER D., "Sensor Network Protocols and Applications", *Proceedings of the First IEEE. International Workshop on In Sensor Network Protocols and Applications*, p. 113-127, 2003.

[KAS 09] KASHIF S., NOSHEILA F., HAFIZAH S., KAMILAH S., ROZEHA R., "Biological inspired self-optimized routing algorithm for wireless sensor networks", *Proceedings of the 9th IEEE Malaysia International Conference on Communications (MICC 2009)*, Kuala Lumpur, 2009.

[KHA 09] KHARGHARIA B., HARIRI S., YOUSIF M., "An Adaptive interleaving technique for memory performance-per-watt maximization", *IEEE Trans. Parallel Distrib. Syst.*, vol. 20, no. 7, p. 1011-1022, July 2009.

[KRI 06] KRIEF F., SALAUN M., *L'autonomie dans les réseaux*, Hermès, Paris, 2006.

[KRI 08] KRIEF F., *Communicating Embedded Systems*, ISTE, London, John Wiley & Sons, New York, 2009.

[LIU 11] LIU X., GHAZISAIDI N., IVANESCU L., KANG R., MAIER, M., "On the tradeoff between energy saving and QoS support for video delivery in EEE-based WiFi networks using real-world traffic traces", *Journal of Lightwave Technology*, 15 September 2011.

[LU 11] LU J., DOUSSON C., KRIEF F., "A self-diagnosis algorithm based on causal graphs", *Proceeding of 7th International Conference on Autonomic and Autonomous Systems (ICAS 2011)*, 2011.

[LYO 08] LYONS M., NEILSON D.T., SALAMON T.R., "Energy efficient strategies for high density telecom applications" , *Workshop on Information, Energy and Environment*, Princeton University, Supelec, Ecole Centrale Paris and Alcatel-Lucent Bell Labs, June 2008.

[MAR 09] MARINKOVIC S.J., POPOVICI E.M., SPAGNOL C., FAUL S., MARNANE W.P., "Energy-efficient low duty cycle MAC protocol for wireless body area networks", *Information Technology in Biomedicine, IEEE Transactions*, vol. 13, no. 6, p. 915-925, November 2009.

[MBA 11] MBAYE M., KHALIFE H., KRIEF F., "Reasoning services for security and energy management in wireless sensor networks", *7th International Conference on Network and Service Management (CNSM 2011)*, Paris, October 2011.

[MOS 10] MOSLEHI K.A., "Reliability perspective of the smart grid, smart grid", *IEEE Transactions*, vol. 1, issue 1, p. 57-64, June 2010.

[MOY 09] MOYA J.M., ARAUJO Á., BANKOVIĆ Z., DE GOYENECHE J.M., VALLEJO J.C., MALAGÓN P., VILLANUEVA D., FRAGA D., ROMERO E., BLESA J., "Improving security for SCADA sensor networks with reputation systems and self-organizing maps", *Sensors*, 2009.

[ONG 11] ONG E.H., MAHATA K., KHAN J.Y., "Energy efficient architecture for green handsets in next generation IP-based wireless networks", *Proceedings of ICC'2011*, p. 1-6, 2011.

[PER 02] PERRIG A., SZEWCZYK R., TYGAR J.D., WEN V., CULLER D.E., "SPINS: security protocols for sensor networks", *Wirel. Netw.*, 8, 5, 521-534, September 2002.

[PÉR 10] PÉREZ-COSTA X., CAMPS-MUR D., "IEEE 802.11 E QoS and power saving features overview and analysis of combined performance", *Wireless Commun.*, 17, 4, p. 88-96, August 2010.

[PER 11] PERES M., CHALOUF M.A., KRIEF F., "On optimizing energy consumption: An adaptative authentication level in wireless sensor networks", *Global Information Infrastructure Symposium*, 2011.

[QIA 05] QIAO D., SHIN K., "Smart power-saving mode for IEEE 802.11 wireless LANs", *24th Annual Joint Conference of the IEEE Computer and Communications Societies, Proceedings IEEE INFOCOM 2005*, vol. 3, p. 1573-1583, March 2005.

[SAK 10] SAKER L., ELAYOUBI S.E., CHAHED T., "Minimizing energy consumption via sleep mode in green base station", *WCNC' IEEE*, p. 1-6, 2010.

[SAL 09] SALEHIE M., TAHVILDARI L., "Self-adaptive software: Landscape and research challenges", *ACM Trans. Autonom. Adapt. Syst.*, 4, 2, Article 14, May 2009.

[SCH 11] SCHOENEN R., BULU G., MIRTAHERI A., YANIKOMEROGLU H., "Green communications by demand shaping and user-in-the-loop tariff-based control", *Online Conference on Green Communications (IEEE GreenCom)*, 2011.

[SHA 07] SHAREE S., LASTER S.S., OLANTUNJI A.O., "Autonomic computing: toward a self-healing system", *American Society for Engineering Education*, 2007.

[SIK 04] SIKORA M., LANEMAN J.N., HAENGGI M., COSTELLO D.J., FUJA T., "On the optimum number of hops in linear wireless networks", *Information Theory Workshop, IEEE*, p. 165-169, 24-29 October 2004.

[SKL 76] SKLAROFF J.R., "Redundancy management technique for space shuttle computers", *IBM J. Res. Dev.*, 20, 1, 20-28 January 1976.

[STE 03] STERRITT R., "Autonomic computing: the natural fusion of soft computing and hard computing", *Proceeding of the IEEE International Conference on Systems, Management and Cybernetics*, vol. 5, p. 4754-4759, 2003.

[WAN 05] WANDER A.S., GURA N., EBERLE H., GUPTA V., CHANG SHANTZ S., "Energy analysis of public-key cryptography for wireless sensor networks", *Proceedings of the Third IEEE International Conference on Pervasive Computing and Communications (PERCOM '05)*, IEEE Computer Society, p. 324-328, Washington DC, United States, 2005.

[WAN 06] WANG Q., HEMPSTEAD M., YANG W., "A realistic power consumption model for wireless sensor network devices", *Sensor and Ad Hoc Communications and Networks, SECON '06. 2006, 3rd Annual IEEE Communications Society on*, vol. 1, p. 286-295, 25-28 September 2006.

[WEN 08] WEN Y.F., CHEN Y.Q., PAN M., "Adaptive ant-based routing in wireless sensor networks using Energy*Delay metrics", *Journal of Zhejiang University SCIENCE A*, vol. 9, p. 531-538, 2008.

[YOO 09] YOO G., LEE E., "Self-healing methodology in ubiquitous sensor network", *School of Information and Communication Engineering*, Sungkyunkwan University, South Korea, 2009.

[ZHE 11] ZHENG X., CAI Y., "Reducing electricity and network cost for online service providers in geographically located internet data centers", *IEEE/ACM International Conference on Green Computing and Communications*, p. 166-169, 2011.

Websites

[4WA] 4WARD FP7 project.

[ANA] ANA, Autonomic Network Architecture FP7 project.

[AST] Tests en environnement radiatif naturel de composants et circuits électroniques: http://www.astep.eu/spip.php?article30.

[AUT] AutoI, Autonomic Internet FP7 project.

[ELE] http://www.electronics-project-design.com/ConformalCoating.html.

[FAR]   http://fr.farnell.com/nichicon/uhd1a471mpd/condensateur-470uf-10v/dp/8822816.

[IEF] https://tools.ietf.org/html/rfc5246.

[LEP]   http://www.lepoint.fr/actualites-sciences-sante/2008-01-03/bugs-celestes/919/0/216898.

[WHI] https://en.wikipedia.org/wiki/Whisker_(metallurgy).

Chapter 7

Reconfigurable Green Terminals: a Step Towards Sustainable Electronics

7.1. Sustainable electronics?

The dark side of Moore's law[1] is the insatiable need of consumers to have the very latest fashionable electronic device. Worldwide, users change their mobile telephones on average every 18 months, because they are encouraged to do so by their service providers [HUA 08] and/or because they want to possess the latest available technology [SAL 08; LI 10a]. Yet the real lifespan of a mobile phone is around 3.5 years [ZAD 10]. Consequently, mobile phones and smartphones have the highest rate of replacement in all of industrial history [ZAD 10].

Chapter written by Lilian Bossuet.

1 This empirical law, or conjecture, pronounced by Gordon Moore in 1975, stated that the number of microprocessor transistors on a silicon chip would double every two years. Generally, this law can be expressed today by stating that year-after-year, the electronic circuits on chips are increasingly powerful, without a concomitant increase in cost.

We can note this trend for the sales of a figurehead product, which very quickly came to occupy a 40% share of the market in smartphones: the Apple *iPhone*. On the day that it was released, over 1.5 million *iPhone* 4G handsets were sold. At that point, it was the fourth version of the *iPhone* in four years since its initial launch, as can be seen from Table 7.1. Apple's commercial strategy is to launch a new version of the product every year, which renders the old version functionally obsolete. The same strategy can be seen for a new product such as the *iPad*, over 300,000 of which were sold on the very first day of its release on the market.

Apple product	Date launched on the market
Original iPhone	June 2007
iPhone 3G	June 2008
iPhone 3GS	June 2009
iPhone 4G	June 2010

Table 7.1. *Evolution of the Apple iPhone range of smartphones in four years*

In general, worldwide in 2009, 1.26 billion mobile telephones were sold (which, for the first time in history, represents a decline in sales of 0.4% in relation to the previous year), and 174.2 million smartphones (which represents a 15% increase in sales in relation to the previous year, and also accounts for the decrease in sales of ordinary mobile phones, due to a shift of buyers' preferences towards these new products).

A similar phenomenon can be seen in the use of computers (portable or otherwise), which have a life expectancy of around 80,000 hours usage, but whose actual lifespan (which corresponds to the duration of usage) is around 20,000 hours [OLI 07].

The high degree of usage of electronic products (computers, communication objects, embedded systems, etc.), coupled with a high rate of replacement (or a reduced actual lifespan) has very serious environmental consequences when taken for the ensemble of these products. This environmental impact is due to many factors. To begin with, the process of manufacturing of these complex products is very energy hungry: it requires numerous materials, chemical products and a great deal of water. Next, the energy consumption during the use of these products may prove significant if we consider the ensemble of the infrastructures (communication infrastructures, for instance). Finally, the manufacturing of the products and their disposal at the end of their lives creates numerous waste products of a greater or lesser degree of toxicity, and which are difficult to deal with. Thus, it is convenient to study the whole of the lifecycle of the electronic products in order to propose the most pertinent solutions possible to reduce their environmental impact.

Of course, in spite of the massive diffusion of electronic products in the world, at present, their environmental impact is not even remotely comparable to that of other industries – particularly transport. For example, the environmental impact of a mobile telephone, from its design to its recycling, is less than that of a family car being driven 100 km [NOK 10]. Another example: in 2004, the company Intel consumed 424 million liters of water a week in the manufacturing of its chips; meanwhile, 757 million liters were needed just to print the Sunday edition of the *New York Times* each week [SHA 04].

These examples might lead one to think that it is unwise to focus on reducing the environmental impact of electronic products. To do so would be to overlook the total impact, both currently and in years to come, which these products have on our contemporary societies. Indeed, electronic products can –

for instance – reduce paper consumption (by way of electronic books and newspapers) and help reduce transport (audio and videoconferences, electronic sending of papers and forms, online administration, etc.). Thus, it is unrealistic for our digital societies not to use these products or to limit their evolution and diffusion. On the contrary, we must be able to continue to develop our society with the help of these products, whilst greatly reducing their environmental impact. Thus, by replacing certain products or improving certain usage, electronic products could, in the future, facilitate a massive reduction in the environmental impact of human activities, if they evolve in a sustainable manner. Indeed, the question of the benefit of technological substitution on the environmental impact of human activities is not clear-cut [FIP 09].

So as to improve the environmental impact of electronic products (for telecommunications terminals, computers, embedded systems, etc.) it is useful to study their lifecycle in depth, from their development and manufacture and until they wind up on the scrapheap. It is a question of carrying out a rapid and simplified *lifecycle assessment* (LCA), such as that offered by the *Centre interuniversitaire de recherche sur le cycle de vie des produits, procédés et services* (CIRAIG[2] – Interuniversity Research Center for the Lifecycle of Products, Processes and Services) [JOL 10]. Such an assessment will clearly reveal which paths ought to be developed in order to reduce that environmental impact which, while negligible in comparison to other industrial products relating to human activities, must be reduced in order to lay the groundwork for a digital society in which electronic products play a very important part. This is what we shall attempt to do in section 7.2, before focusing on a new avenue: the design of a reconfigurable hardware system in order to increase the lifespan of electronic products by reducing their functional obsolescence.

---

2 www.ciraig.com.

7.2. Environmental impact of electronic products during their lifecycle

7.2.1. *Lifecycle of electronic products*

Electronic products follow a classic lifecycle, as Figure 7.1 shows in a simplified manner (this figure is inspired by that given by Dhingra in a 2010 article [DHI 10], but is more complete in terms of the risks associated with each stage of the lifecycle). The four essential phases of the lifecycle of such products are the processing of raw materials, manufacturing of the products, their usage and finally their end of life (recycling and disposal). These phases do not have the same environmental impact, in volume and in consequence. The lower level in Figure 7.1 reviews some of the risks incurred during each of these phases. Impacts on the health and safety of workers and on local residents from discharges.

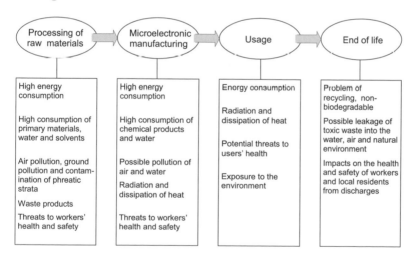

Figure 7.1. *Lifecycle of electronic products and main risks associated with each stage*

Figure 7.2 takes the example of a mobile telephone from the company Nokia [NOK 10], and gives the proportion of each of the lifecycle phases in the use of energy and greenhouse gas emission. The percentages given are for an actual lifespan of 36 months. What becomes clear is that the phases of creation of the product, going from the primary and secondary materials to the manufacture of the final product, have the greatest environmental impact. By contrast, the phase of actual usage of the product (for 36 months) represents just slightly more than a quarter of the energy consumption and greenhouse gas emission. This proportion decreases greatly with a shorter lifespan – for instance, for a lifespan equal to the average life of mobile telephones (18 months [HUA 08]), it drops to 16.4% in terms of energy consumption. This can be explained simply by the fact that the other phases of the product's lifecycle have a fixed environmental impact which is independent of the actual lifespan of the product.

Figure 7.3 again uses the data from Figure 7.2, but shortens the usage from 36 to 18 months. In this case, we note that the energy needed for the use of the telephone throughout its operational life is roughly the same as that needed for transport (of the primary material and of the finished product). Thus, we can easily see why it is essential to increase this lifespan and to counter an obsolescence accelerated by the arrival of new products (see Table 7.1). However, we must not forget that the energy really used during the actual lifespan of the telephone for its main function is in fact divided across other products which constitute the infrastructure of the communication network (such as the base transceiver stations (BTSs) and the servers). Some studies show that the consumption of the infrastructure represents 90% of the total consumption needed for wireless communications [FLI 08]. While, in the case of mobile telephones and computers (either desktop or laptop), the energy consumption and greenhouse gas emissions are greater during production than during usage, the same is not true of servers [FLI 09].

In the case of Figures 7.2 and 7.3, the recycling of the telephone represents only a very small (negligible) percentage of the energy consumption and greenhouse gas emissions. However, less than 5% of used mobile telephones are recycled [FLI 09], which might account for the negligible contribution of telephone recycling to their environmental impact.

The rest of this section goes into greater detail about the environmental impact of the main phases of the lifecycle of electronic products.

7.2.2. Microelectronic manufacturing

The microelectronics industry is a very heavy consumer of primary materials, chemical products, water and energy (electric, gas and fossil fuel). As we saw earlier, more than half of the energy consumed during the course of a mobile telephone's life is attributable to the process of manufacture (mainly of the electronic components). However, the manufacture of electronic components is highly complex. For the latest technologies, the manufacture of an electronic chip requires between 400 and 500 stages [BRA 08; BRA 10]. The main stages of manufacturing of microelectronic products are: the development of the product (very often overlooked in studies estimating energy usage); extraction of the primary materials, treatment of those materials (purification of silicon and water); production of chemicals (of which, as we shall see later, there are many); procedures of chemical and optical treatments (amongst others: silicon doping, lithography, epitaxy, diffusion, etc.); testing and cutting of the wafers[3]; bonding; and packaging.

---

3 For the microelectronics industry, *wafers* are slivers of a silicon bar, currently 300 mm in diameter and of sub-millimetric thickness, on which the electronic circuits are etched. Depending on their size, it is possible to etch between a few dozen and several hundred circuits on a single wafer.

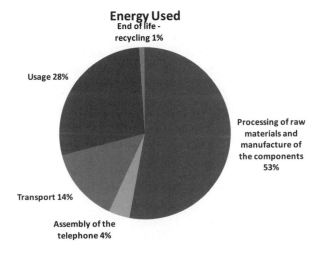

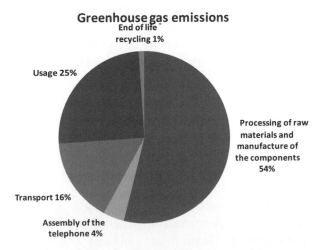

Figure 7.2. *Contribution of the different stages of the lifecycle of a Nokia mobile telephone in energy consumption and greenhouse gas emissions for a usage lifespan of 36 months [NOK 10]*

Once the chip is in its package, numerous other stages are necessary in order to get a finished electronic product:

amongst others, the creation of the printed circuit board (PCB) which accommodates the different integral components and enables us to link them via copper tracks. Again, test phases are necessary for the PCB.

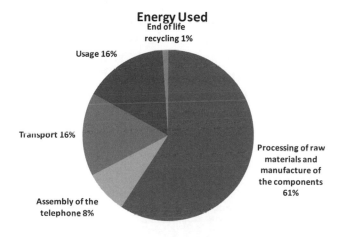

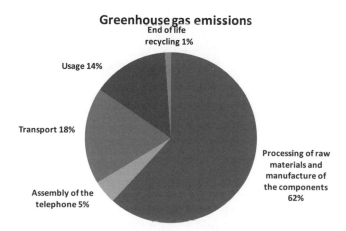

Figure 7.3. *Contribution of the different stages of the lifecycle of a Nokia mobile telephone in energy consumption and greenhouse gas emissions for a usage lifespan of 18 months*

In 2002, a very exhaustive study about the quantity of materials, gas and water needed to construct an electronic circuit was published [WIL 02]. The authors of this study show that the manufacture of a 32-megabyte memory, of DRAM[4] type in CMOS[5] technology – the weight of which is 2 grams – requires 72 grams of chemical products and 1.6 kg of fossil fuel for energy. Thus, the weight of the secondary materials (in comparison to silicon, which is the primary material) is around 630 times that of the end product. Figure 7.4 is a diagrammatic representation of the low mass yield during the manufacture of a wafer 150 mm in diameter [WIL 02]. This yield tends to increase with technologies but to decrease with the size of the chips [OLI 07].

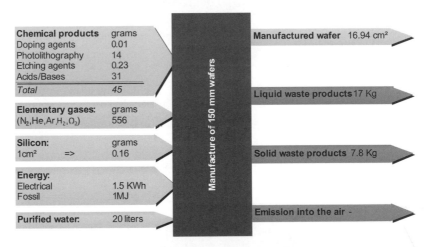

Figure 7.4. *A graphic showing the masses of input and output in the process of manufacturing a wafer 150 mm in diameter [WIL 02]*

---

4 *Dynamic Random Access Memory*, a very small volatile memory (only one MOS transistor), the value of which constantly needs to be refreshed dynamically.

5 *Complementary Metal Oxide Semiconductor*, the most widely used microelectronic technology, based on using a pair of MOSFET transistors in the switching regime. In all CMOS circuits, each N-MOS transistor is systematically and symmetrically associated with a complementary P-MOS transistor. When one of these transistors is blocked, the complementary transistor is passing.

The manufacture of microelectronic components uses a wide variety of chemical substances. Many of these are toxic and could potentially have a considerable impact through emissions of polluted air and water. These chemical components are mainly used in silicon doping, in the etching of metallic connections (made of copper until the 2000s, and aluminum since then) and for photolithography.

In 2002, the manufacture of a 1 cm^2 electronic circuit required 45 grams of chemical products, which represents a ratio of 280 kg of chemical products to 1 kg of silicon [WIL 02]. The use of a metric such as weight is questionable, because it is heavily dependent on the technology used, the size of the wafers (for instance, before 2000, their diameter was 150 or 200 mm; since 2001 it has been 300 mm, and in 2012 it will be 450 mm) and the number of metallic layers [SHA 04].

Nevertheless, the weight of the material used in the manufacturing process is an interesting criterion to use, because it allows us to quantify the mass yield of microelectronic manufacturing [WIL 04]. In addition, analysis of the mass yield is commonly used under the name *ecological rucksack* [FLI 09].

In order to complete this point about the use of chemical products, Table 7.2 lists and classifies the main chemical agents used by the microelectronics industry, such as elementary gases, doping agents, etching agents, acids, bases and photolithography agents.

The quantity of water used in the manufacturing of an electronic chip is also considerable. A microelectronics factory uses around 10 million liters of water a month to manufacture tens of thousands of wafers. In 2002, for instance, this corresponded to a consumption of 18 to 27 liters of water per cm^2 of wafer manufactured [WIL 02]. This amount of water consumed varies greatly from one technology to another. However, in any case, the microelectronics industry needs extremely pure water (at least a million times purer than tapwater). In addition, the purification of the water is normally done as close as possible to its usage, in order to limit any contamination or deteriorations in the purity, which necessitates a great deal more energy on the microelectronic production site.

Type of chemical agent	Main agents used
Elementary gases	Helium (He$_2$), dinitrogen (N$_2$), dioxygen (O$_2$), argon (Ar), dihydrogen (H$_2$)
Doping agents	Silane (SiH$_4$), phosphorus pentoxide (P$_2$O$_5$), phosphorus oxychloride (POCl$_3$), phosphorus hydride (PH$_3$), diborane (B$_2$H$_6$), arsine (AsH$_3$), dichlorosilane (SiH$_2$Cl$_2$)
Etching agents	Ammonia (NH$_3$), nitrogen protoxide (N O), dichlorine (Cl$_2$), boron trichloride (BCl$_3$), boron trifluoride (BF$_3$), hydrogen bromide (HBr), hydrogen chloride (HCl), hydrogen fluoride (HF), nitrogen trifluoride (NF$_3$), tungsten hexafluoride (WF$_6$), sulfur hexafluoride (SF$_6$), methane trifluoride (CHF$_3$), carbon tetrafluoride (CF$_4$)
Acids/Bases	Hydrofluoric acid (HF), ammonium (NH$_4^+$), acid (H$_3$PO$_4$), nitric acid (HNO$_3$), sulfuric acid (H$_2$SO$_4$), hydrochloric acid (H$_3$O$^+$Cl$^-$), ammonia (NH$_3$), hydrogen chloride (HCl), sodium hydroxide (NaOH)
Photolithography agents	Hydrogen dioxide (H$_2$O$_2$), isopropyl alcohol (CH$_3$CH(OH)-CH$_3$), acetone (CH$_3$COCH$_3$), tetramethylammonium hydroxide ((CH$_3$)$_4$NOH)

Table 7.2. *Description of the main chemical agents used in the manufacturing of a microelectronic chip in CMOS technology [WIL 02]*

Other materials are needed for the manufacture of the integrated circuit (the chip in its casing) such as ceramics, plastic, gold, nickel, copper and/or aluminum. The manufacture of the printed circuit board (or electronic card) also uses a great many materials such as epoxy, copper, gold and tin for soldering.

According to Williams [WIL 02], during the manufacture of a microelectronic chip, 83% of the energy used is electric; the rest of the energy is sourced from petroleum, gas and kerosene. The microelectronics industry is therefore a huge consumer of electricity. For instance, between 1998 and 2002, even though the industry was massively concentrated in Asia, it accounted for 1.5% of industrial electricity consumption in the United States [BRA 10].

The electrical consumption needed for the manufacture of a chip depends greatly on the technology and on the diameter of the wafers, as shown in Table 7.3, which gives the result of several studies about the electrical energy consumption of the microelectronics industry for the manufacture of a wafer between 1993 and 2008 [DUQ 10].

The energy consumed for manufacturing wafers stems from the consumption of the many systems necessary to carry out the procedures. Figure 7.5 shows the proportion of the most energy-hungry systems during the manufacture of microelectronic circuits [BRA 10]. It is indeed the systems needed for the microelectronic manufacture (diffusion, photolithography, plating, etc.) which consume the most energy, followed by the cooling system.

Ultimately, the manufacture of a wafer requires a considerable amount of energy: a figure of 2 gigajoules is given for a wafer 300 mm in diameter [OLI 07], which corresponds to around 45 liters of petroleum. By way of comparison, an Intel® Core™ 2 Duo processor (45 nm, 2.53 GHz) must be used for over 10 years over 80% of the time (300 days per year) for four hours a day in order for its energy consumption to be equivalent to that needed for its manufacture. Bear in mind that in order for the energy used in the manufacture of an electronic circuit to be negligible (less than 10%) in relation to its consumption in usage, it would have to be used for 100 years!

Literature	Year	Diameter of wafer	Electricity requirements
[MCC 93]	1993	150 mm	285 kWh/wafer
[MUR 01]	2001	200 mm	440 kWh/wafer
[KUE 03]	2003	200 mm	499 kWh/wafer
[MUR 03]	2003	300 mm	664 kWh/wafer
[KEM 05]	2005	300 mm	583 kWh/wafer
[KRI 08]	2008	200 mm	470 kWh/wafer

Table 7.3. *Electrical energy needed to manufacture a wafer, given by several studies between 1993 and 2008 [DUQ 10]*

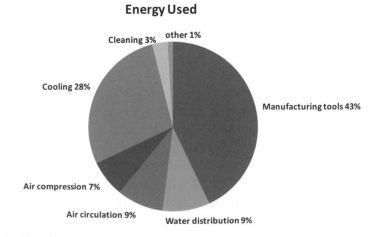

Energy Used

Cleaning 3%   other 1%

Cooling 28%

Manufacturing tools 43%

Air compression 7%

Air circulation 9%   Water distribution 9%

Figure 7.5. *Proportion of energy consumption of different systems needed to carry out the process of microelectronic manufacture [BRA 10]*

7.2.3. *Usage of electronic products*

As we saw in the introduction to this chapter, the energy consumed during the actual operational lifespan of an electronic product represents a small proportion of the total energy required for the entire lifecycle of the product. We have also seen that the amount of energy consumed at the "usage" stage is proportional to the lifespan. A very short lifespan makes this contribution to the total energy consumption almost negligible (see Figure 7.3). This remark makes R&D projects seeking to reduce the power and energy consumption of electronic products during their operation seem petty and trivial.

Indeed, projects which would yield, say, a 10% reduction in the energy consumption of a mobile telephone (which is difficult to achieve) throughout an operational lifespan of 18 months would only yield a reduction of around 1.5% in the total energy consumption throughout the whole lifecycle of the telephone. Paradoxically, this is a sizeable field of academic research, which involves a large community, numerous conferences and specialized journals. Does this mean that such works are not highly useful? No, it does not – by reducing the power and energy consumption of electronic products, what is at stake is the consumption due to their supply. This is particularly important for products supplied by batteries. Reducing the power consumption facilitates a decrease in the size and weight of the batteries. Reducing the energy consumption in turn facilitates an increase in the autonomy of the electronic products.

Another very important point: power and energy consumption is linked to the heat dissipation from the electronic chips; yet this becomes prohibitive (because of the size and complexity of the cooling systems needed) without control of consumption. Thus, we can say from this brief analysis that the design of low-energy electronic products is an important field of research, but not that it alone does not constitute the only path towards green and/or durable electronics.

In addition to the work aimed at directly reducing the consumption of electronic circuits in operation, certain projects, at the level of applications, seek to involve the user in managing the energy consumption of the product. For instance, that is the purpose of the proposed "green switch" for mobile telephones [ZAD 10]. With the green switch, the user is aware of the energy needs of his device, and can choose between higher performance and lower energy consumption. This type of application is aimed at modifying usage, and making the user an *energy-conscious user* in relation to his device and its usage. In a future where the cost of energy could increase dramatically, this type of usage could be generalized and become absolutely necessary.

7.2.4. *Electronic waste products*

Once electronic products are thrown out, their life does not stop, because a final cycle commences as electronic waste products. These are constantly evolving. According to a study by the European Union, they are increasing by 3-5% a year, which corresponds to around three times the rate of increase in other household waste [SCH 05]. From an international point of view, this is the flow of waste which is increasing most at present. The volume of electronic waste products is around 50 million tons per year on a global scale. By way of example, in 2005, 130 million mobile telephones were thrown away, which represents 65,000 tons of waste. In Europe (the 25 member state EU) in the same year, the total volume of electronic waste represented 8 kg per person, equating to 3.6 million tons of waste.

This mass of waste causes serious problems. First of all, these are not harmless waste products: the materials and chemical components which go into making electronic products mean they produce highly dangerous and toxic waste [HER 07]. It is complicated and costly to recycle them: only around 17% of electronic waste is actually recycled [LAS 10]. This has led to the emergence of more-or-less "official" treatment facilities in Asia and Africa [HUA 09; EUG 08].

The current situation is untenable in the long term; therefore it is necessary to find solutions which help to reduce this waste. One of the most widely accepted paths today is to follow the *3R*s; *Reduce, Reuse* and *Recycle*. However, it is possible to envisage adding a fourth *R* to this rule: *Reconfigure*. We shall explore this avenue in the rest of this chapter.

7.3. Reduce, reuse, recycle and reconfigure

7.3.1. *Reduce, reuse, recycle*

The first term of the *3R*s rule corresponds to reduction. This has to be a guiding principle for all stages of the lifecycle of the product. At the manufacturing stage, it is necessary to reduce the quantity of materials, chemical products, water and energy that are needed. In this domain, numerous projects are underway, and are encouraged by the fact that these reductions may also entail a reduction in production costs. If so, then industrial players will implement the principles of reduction. During the usage phase, reduction is paradoxically to be seen in an increase: an increase in lifespan. We saw earlier that this increase leads to a better ratio between the energy "usefully" consumed and that needed for the manufacture of the product. The consumption of power and energy during operation can also be reduced. This inevitably involves making better use of the electronic products, e.g. charging them only when absolutely necessary. Finally, an overall reduction in waste must be envisaged at every stage, beginning with the manufacturing of the products. To this end, legal incentives, such as those already in existence, must create a link between the manufacturer of the product and that product as waste. It must be noted that, in general, in order to actually reduce, we must be able to correctly estimate. However, it often proves difficult to obtain precise estimations of the environmental impact of a product during its entire lifecycle [DUQ 10; NIG 10; COR 10].

The second point of the *3Rs* rule corresponds to reuse. To illustrate this approach in the context of microprocessors, a 2007 publication [OLI 07] proposes that there be a "food chain" of microprocessors, which would compensate the energy needed to manufacture the microchips over several generations of products. This chain is imagined as follows: a new generation of microprocessor replaces an existing microprocessor in a top-of-the-range device (a powerful laptop, for example); this existing processor is not thrown away, but rather replaces a less powerful microprocessor in an existing device (say, a PDA), which then replaces an older generation of processor in another device (e.g. a games console). Figure 7.6 illustrates the process of reuse in terms of microprocessors.

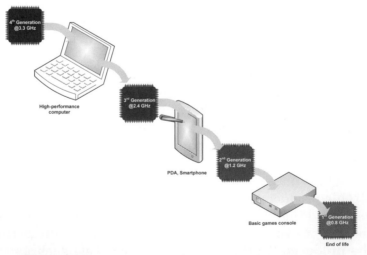

Figure 7.6. *Illustration of a "food chain" of microprocessors*

In theory, this solution is attractive. Indeed, each one of the devices can obtain a gain in power without this involving the manufacture of a great many chips. However, in reality, this method of hardware updating by reuse of circuits is quite simply not feasible. Firstly, from a functional point of view, each circuit, each microprocessor, is fairly specialized, or at least has inherent specificities which make functional compatibility difficult. This point could be dealt with at the level of software applications, on condition that a certain form of virtualization is used. However, what limits reuse more definitely is the physical specificity of the circuits (dimensions of integrated circuits and number of input/output pins) and the levels of voltage and current acceptable on the power supply and on the input-output diagrams.

Thus, the reuse of electronic circuits appears to be complicated; yet they can be reused between electronic devices such as mobile telephones or smartphones [LI 10a]. For instance, a recent study shows how old smartphones (which very rapidly become obsolete) can be used in school programs for didactic software applications requiring a limited number of hardware resources which are usually found in these devices [LI 10b].

The idea of updating hardware is a good one, because it helps combat too rapid an obsolescence of the hardware, and increases the functional lifespan of electronic products. Thus, it is a question of offering updated hardware, the same way as we do for software (imagine how the computer market would be if, every time a new version of a software package was released, you had to replace your computer!). The possibility of updating the hardware came into being at the start of the 2000s, with new reconfigurable hardware circuits – FPGAs (*Field Programmable Gate Arrays*) – and more generally, so-called reconfigurable architectures [BOS 10; BOS 06]. We shall discuss this point in section 7.3.2.

To conclude, as we saw earlier (section 7.2.4), electronic waste products are very difficult to recycle. This last part of the *3R*s rule seems to be the most complicated [HUA 09; EUG 08; HER 07]. However, this avenue certainly offers a great deal of room for progress, which would enable us to achieve better recycling [LAS 10].

7.3.2. *Reconfiguring with the help of FPGAs*

The aim of this section is to explain to non-expert readers what exactly an FPGA-type reconfigurable hardware circuit is. Experienced readers can skip directly to section 7.4, which presents examples of the use of FPGAs in reconfigurable terminals with prolonged actual operational life.

FPGAs are configurable hardware circuits for digital electronics [BOS 10; MAX 04]. In their initial state, they can do nothing, but contain a large amount (depending on the technology used) of operational hardware resources whose function we can configure. These resources are, primarily, logic elementary blocks (to generate Boolean functions), RAM memories, fixed-point arithmetic operators, internal routing resources and input/output resources. These configurable resources are linked by a dense network of data routing lines and clock signal routing lines. These routing lines are also configurable.

In addition to these resources, an FPGA contains an internal configuration memory. Each point in this memory corresponds to the configuration of an element of one of the operational resources. In most cases, this memory is created using one of the following three technologies:

– Antifuse (the oldest, which can only be configured once);

– Flash (non-volatile);

– SRAM (volatile, the most widely used, representing over 80% of the market).

As Figure 7.7 shows, in order to create an application with an FPGA, we have to describe the electronic circuit to be created using a hardware description language such as VHDL[6] (*Very High Speed Integrated Circuit Hardware Description Language*). We then have to synthesize that description in an electronic circuit. This and the following stages can be performed using free software supplied by the circuit manufacturer. Finally, following a stage of placement and routing which takes account of the architecture of the FPGA, a configuration file called *bitstream* is generated. This allows us to specify the position of the points in the configuration memory when configuring the FPGA.

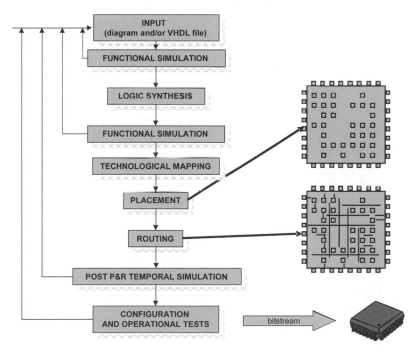

Figure 7.7. *Conventional flow of FPGA conception (simplified)*

---

6 Other languages such as Verilog and SystemC do exist, but VHDL is still currently the most widely used in Europe.

Of the main manufacturers of FPGAs around the world, we can cite: Xilinx (1st in the FPGA market, 53% market share in 2009) [XIL 10]; Altera (2nd in the FPGA market, 36% market share in 2009); [ALT 10], Microsemi (1st in the Antifuse and Flash FPGA market) [ACT 10]; Atmel; QuickLogic; Lattice; and M2000 (FPGA cores).

FPGAs are the very latest in digital circuits; consequently they are still in the full throes of development. Their architecture has evolved over the past few years, as has the granularity and the type of configurable logic resources. One of the most interesting evolutions relates to their usage. Partial reconfiguration and dynamic reconfiguration have opened up a new field of application for these circuits, which are occupying an ever-increasing share of the market of hardware digital circuits (excluding microprocessors), which is dominated by ASICs[7] (*Application Specific Integrated Circuits*).

Indeed, in an uncertain global economic climate, FPGAs seem a flexible solution, well adapted to economic constraints such as the reduced time to market and the potential for evolution or flexibility of the products. In addition, the economic model associated with FPGAs, which is a linear model, is becoming increasingly advantageous in comparison with the economic model of ASICs, for which the cost of producing the first prototype means it would take a very long time to offset the cost of this solution, and would only be economically viable for production on a very large scale. Figure 7.8 shows that the number of circuits manufactured for which the ASIC solution is economically more profitable (the crossover point) tends to increase with the evolution of the technologies. For instance, with the technology used in 2003 (90 nm), the ASIC solution becomes interesting from about when a million circuits are to be made

---

7 An ASIC is an electronic circuit *made to measure*. These are the highest-performing circuits but are also the most costly, and take longest to make.

(and therefore sold). Hence, FPGA solutions are increasingly advantageous both from a technological and an economic standpoint.

Since the 2000s, the density of integration of FPGA circuits means that on a single chip, we can assemble a matrix of configurable hardware elements (logics, memories, arithmetic operators, input-outputs), and one or more microprocessors. This type of circuit enables us to take advantage of the parallelism of calculation offered by the hardware architecture and of the efficient sequential control offered by the programmable system (microprocessor). By taking advantage of the respective properties of programmable systems and reconfigurable systems, it is possible to improve the appropriateness of the overall system for the application being developed. In this case, the use of joint hardware/software design techniques is essential, and requires a great deal of effort in development of tools [CTI 98].

Today, there are various hybrid architectures in existence. Figure 7.9 illustrates these different possibilities. In certain circuits, the configurable hardware part and the programmable part are separated by a specific bus. The programmable part includes the entirety of the microprocessor system: processor core, cache memories, peripheral memories, interfaces, etc. This was the case for the first commercial circuit to include a processor core, the Altera Excalibur circuit, which contained an APEX™ 20KE FPGA and a 32-bit ARM9 processor core, functioning at 100 MHz, along with two sets of 8-kilobyte cache memory (instructions + data). Unfortunately, at the time that this component was released, the assisted design tools were not at a sufficient level of maturity to facilitate its simple and efficient use in an industrial context.

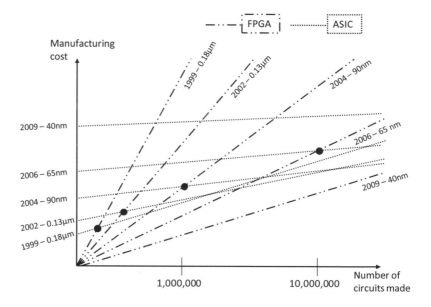

Figure 7.8. *Models of costs for ASIC and FGPA components (SRAM)
depending on the technology used*

In certain circuits, the processor(s) are embedded in the heart of the configurable matrix. The processors are not necessarily accompanied by their system, but rather by software tools which can be used to configure part of the logic to constitute the whole system. This offers greater flexibility in the choice of programmable system. It is this solution which was chosen by Xilinx for its first hybrid component, Virtex-II Pro. This was made up of a Virtex-II Pro matrix and between one and four 32-bit IBM PowerPC 405 cores, calibrated at 400 MHz, with two sets of 16-kilobyte cache memory (instructions + data). This same architecture is used even today for the very latest generation Virtex components.

It should be noted that it is possible to create a hybrid circuit without physically having a processor core embedded in the FPGA circuit. In this case, the use of a synthesizable core (called a *soft core*), provided for free by the circuit manufacturer, is a very effective solution. For instance, the Altera NIOS [ALT 09] and Xilinx MicroBlaze [SUN 09] 32-bit synthesizable cores are very widely used. Of course, the performances of such cores are inferior to those of embedded cores, but they do offer greater flexibility of configuration.

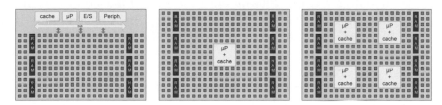

Figure 7.9. *Three possible architectures for hybrid-circuit FPGA-microprocessor(s)*

The configuration of FPGAs may take different forms, whether it be to configure the operating resources or to configure the routing networks. However, FPGAs using antifuse technology can only be configured once: we speak of OTP (*One Time Programmable*) circuits. Note the use of the word "programmable", although FPGA circuits are not programmable (they do not execute a program) but rather configurable (the memorized configuration of all the functional and routing elements of the FPGA enables it to carry out an application). The only FPGAs which are reconfigurable (configurable on several occasions) are circuits using Flash technology (non-volatile memory) and those using SRAM technology (volatile memory which necessitates that the configuration be saved in an external, non-volatile memory).

In these latter two cases, the reconfiguration of the architecture can be achieved in various ways during the execution of the application. This can happen only once without causing difficulties during the execution of the application. In this case, we shall speak of static reconfiguration. The processes of reconfiguration and execution of the application are, in this case, distinct and clearly separate in time. A new reconfiguration will be performed in both cases. To begin with, this may be the result of loss of the configuration – which, for certain devices, may be due to cutoff in the power supply (this is the case for FPGAs using SRAM technology). It may also come in the wake of a modification of the application by the designer if that application is defective or can be improved (this is the case of prototyping).

However, an in-depth study of the execution of an application may reveal that certain configured parts of the architecture are only necessary for a short time in relation to the duration of execution, but can occupy a significant amount of space in the architecture. The ratio between the time of usage of the configured part and the space it takes up in the architecture may be small – hence the idea of introducing temporal dynamism into the reconfiguration. When part of the application has been executed, and in the case where it will not be executed again for some time, we can reconfigure the elements which were devoted to it and use them for another part of the application. This reconfiguration is performed in parallel with the execution of the application.

Thus, dynamic reconfiguration enables us to optimize the configured surface over time. Given that it is a question of modifying only a part of the configurable elements in real time, we have to use a partial-reconfiguration architecture. The disadvantage to this is that we have to clearly determine the partitioning of the application over time in order to take advantage of the whole surface of the circuit. Fragmentation issues may arise, as happens on computer hard drives [COM 99]. Furthermore, we have to correctly establish communications between the partitions [DEL 07].

Figure 7.10 is a diagrammatic representation of the progressing of reconfiguration during the execution of the application in both cases: complete reconfiguration and dynamic partial reconfiguration. During the run time of the application, complete reconfiguration is static, in contrast to partial reconfiguration. The latter, in certain conditions, enables us to reconfigure part of the circuit during operation without affecting the configured parts which must not be modified. In this case, we speak of dynamic configuration.

For each of the cases in Figure 7.10, a gray square represents a configured element, and a white square represents an unused element. Where two squares have different levels of gray, this means that they are used to carry out two distinct parts (or tasks) of the application (particularly in the case of partial reconfiguration).

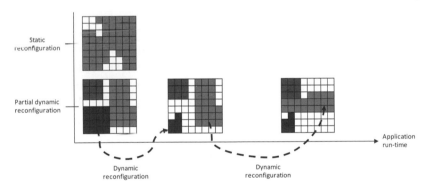

Figure 7.10. *Static reconfiguration and partial dynamic reconfiguration during run-time of the application [BOS 10]*

In any case, managing the configuration – particularly with the notion of dynamic reconfiguration – is a complex task. More often than not, an auxiliary circuit has to be installed in order to manage it. This circuit may be a processor either external or internal to the architecture. This latter solution is offered by Xilinx in their self-reconfiguring system, based on the use of a MicroBlaze onboard processor [BLO 03; ULM 04].

Thanks to its capacities for configuration, (dynamic) self-configuration and partial reconfiguration, FPGA is an ideal candidate to update hardware and thereby reduce the functional obsolescence of a great many electronic products. Thus, *Reuse* from the *3R*s is supplemented by *Reconfigure*. However, we have to develop systems and architectures which are reconfigurable so that they can actually evolve over the course of time.

7.4. Examples of reconfigurable terminals

In recent years, we have witnessed a spectacular boom in communicating systems (systems for processing and exchange of information, telecommunications systems, surveillance and monitoring systems, etc.), which involves a multitude of machines and sometimes functional redundancy. In addition, for each of these systems, the evolutions are rapid; as soon as a new communication standard or a new service appears, we observe an accelerated obsolescence which is not physically justified (the products are "not technologically worn out"). The consequences for the environment are very real and dire. The main concern is to come up with a simple response to the question "What are we to do if a new service or a new standard appears?" One envisageable response is to update the software and the hardware, by way of mass use of heterogeneous reconfigurable architectures.

This is particularly interesting when the electronic circuitry is embedded in systems with a long lifespan: habitat (domestic home, furniture, etc.), land vehicles (trains, trucks, cars, etc.), aeronautics (satellites, airplanes, etc.), industry (power plant, production, etc.). From a societal point of view, this leads to a decrease in the costs of changing communication standards, an increase in the flexibility for evolution of services, an increase in the time-to-market (of the hardware and of the service), a reduction in technological waste because of an extended lifespan (use) of telecommunications systems, and finally a reduction in technological obsolescence. Thus, the use of reconfigurable heterogeneous architectures takes its place in the context of sustainable development. In this context, the aim is to offer *more* services with *fewer* pieces of new hardware and energy resources.

Figure 7.11 illustrates the possible industrial environment of a reconfigurable communication gateway, called a *Chameleon Gateway*, which can be reconfigured and updated to be compatible with new communication standards. This architecture has been partially implemented in the context of a sensor network by implanting the AODV [MAG 08] and OLSR [RIB 09] algorithms. Figure 7.12 illustrates the internal architecture of the chameleon gateway, centered on an FPGA-type reconfigurable hardware core.

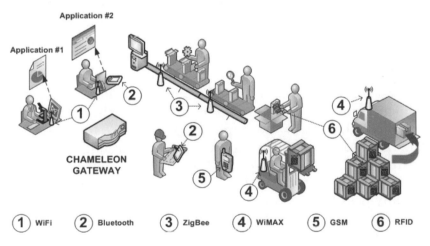

Figure 7.11. *Industrial environment of a chameleon gateway*

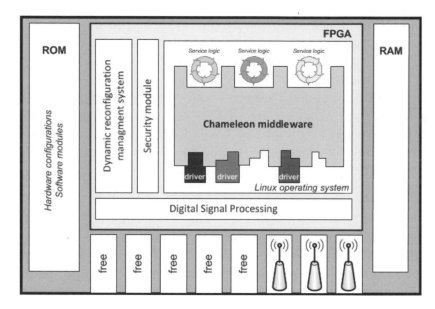

Figure 7.12. *Internal architecture of the chameleon gateway*

So as to eliminate problems related to the reuse of microprocessors, which we discussed above, another study instead suggests using FPGAs to create reusable electronic systems [LEH 10]. The concept of a reusable electronic system championed by this study contains two parts: a reconfigurable FPGA to carry out the treatments inherent to the application and a chip-mounted system called FPESS (*Field-Programmable Electronics Support System*) which serves as a configurable interface with the sensors and triggers inherent to the application. Figure 7.13 illustrates the architecture of the concept of a reusable electronic system [LEH 10]. In this context, and in contrast to the chameleon gateway, partial reconfiguration is not involved for any possible hardware update – in this case, rather, it is a question of using reconfiguration in order to adapt to several applications.

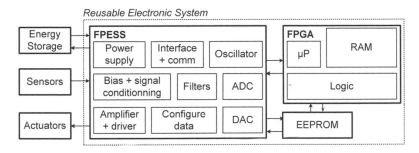

Figure 7.13. *Architecture of the concept of a reusable electronic system [LEH 10]*

From an applicative point of view, reconfiguration can be thought of at several different levels so as to increase the flexibility and adaptation of the system. For instance, in the case of software-defined radio (SDR), reconfiguration can intervene both at the level of wave-shape processing [DEL 07] and at the level of the front-end radiofrequency [DEJ 07]. That said, beyond the aspects of adaptation and updating, reconfiguration of the hardware is essential for adequately managing the energy and the power consumption of modern communications systems [DEJ 07].

The most common objection raised about updating hardware in order to prolong the operational life of electronic products is that of the economic context. The holders of this view predict that the manufacturers of electronic products will take a dim view of technological solutions which aim to reduce their sales figures [FIP 09] – obviously, one of the consequences of an increase in the lifespan of electronic products will be a reduction in the sales of such products. However, one can easily look at the world of computing to see whether updating of software has led to a decrease in that sector. The answer is, categorically, that it has not – on the contrary, it has led to impressive economic dynamism, driven by sales of software updates. As regards electronic circuits, such a strategy would lead to the emergence of a new sector of the market: that of sales of hardware updates. In a manner of speaking, it is a question of de-materializing the economy of electronic products, by proposing to substitute the sale of services for that of goods. Also, using reconfigurable technology, industries could reduce the time-to-market by more rapidly launching products with an initial version whose performances/functionalities can be improved by updates. Thus, it is a question of encouraging the emergence of a function-based economy.

In order for this concept to develop, efforts are needed in terms of standardization of reconfiguration. Today, every manufacturer of FPGAs holds the rights on their own systems for configuration. One solution in order to overcome the technological issues may stem from the implementation of the concept of virtualization for hardware systems. A virtual (open or otherwise) reconfigurable FPGA (be it partially or completely reconfigurable) could be used by developers of hardware updates and be distributed more easily. Today, this concept may appear unrealistic, but what will become of it in a future where the cost of primary materials, energy and manufacturing needed for electronic products becomes prohibitive?

7.5. Conclusion

In this chapter, we have been able to make the observation that the electronics industry today is not yet *green* and/or *sustainable*, which are not the same thing. In order to attempt to achieve this objective, we have suggested a new approach, which consists of reducing the functional obsolescence of electronic products by updating the hardware, based on the reconfiguration capacities of reconfigurable circuits such as FPGAs. However, we are not of the opinion that this solution alone will suffice to offset the environmental impact of electronic products. A certain number of points must be explored in conjunction in order to make real progress towards a more sustainable, or *greener*, electronics industry. Solutions exist at the level of technology, such as the use of carbon nanotubes or graphene. At the level of the manufacture of microchips, new assembly methods, such as 3D construction of the integrated circuits, seem good solutions to reduce the energy needed for microelectronics production [WAN 10]. At the level of the system, the reduction in power consumption and energy consumption remains an important objective to achieve. However, we must also think about making systems more flexible, more homogeneous, in order to facilitate reuse and reconfiguration. The principles of virtualization, extended to the ensemble of components of the system, may offer interesting solutions. The design of these systems must be guided by new constraints which take account of the environmental impact of the system under development. This will involve defining new metrics which go beyond the simple aspect of the power consumption. In conclusion, such approaches must rely on the development of engineer training programs in order to generalize and improve practices.

7.6. Bibliography

[ACT 10] ACTEL COPORATION, www.actel.com, 2010.

[ALT 10] ALTERA COPORATION, http://www.altera.com, 2010.

[ALT 09] ALTERA COPORATION, NIOS II Processor Reference Handbook, version 9.1, November 2009.

[BLO 03] BLODGET B., JAMES-ROXBY P., KELLER E., McMILLAN S., SUNDARARAJAN P., "A self-reconfiguration platform", *Proceedings of 13th International Conference on Field-Programmable Logic and Applications (FPL 2003)*, Lecture Notes in Computer Science, vol. 2778/2003, p. 565-574, 2003.

[BOS 06] Bossuet L., Gogniat G., Philippe J.L., "Exploration de l'espace de conception des architectures reconfigurables", *Revue des techniques et sciences informatiques, série TSI, Architecture des ordinateurs*, vol. 25, no. 7/2006, p. 921-946, 2006.

[BOS 10] Bossuet L., *Les architectures matérielles reconfigurables. De la modélisation à l'exploration architecturale*, Editions universitaires européennes, Saarbrücken, Germany, 2010.

[BRA 08] Branham M.S., Semiconductors and sustainability: energy and materials use in integrated circuit manufacturing, Masters Thesis, department of Mechanical Engineering, Massachusetts Institute of Technology, Cambridge, MA, United States, 2008.

[BRA 10] Branham M.S., Gutowski T.G., "Deconstructing energy use in microelectronics manufacturing: an experimental case study of a MEMS fabrication facility", *Environmental Science & Technology*, ACS, vol. 44, no. 11, p. 4295-4301, 2010.

[COM 99] Compton K., Programming architectures for run-time reconfigurable systems. Masters Thesis, Dept of ECE, Northwestern University, Evanston, Illinois, United States, December 1999.

[COR 10] Corrigan K., Shah A., Patel C., "Estimating environmental costs", *Proceedings of the First USENIX Workshop on Sustainable Information Technology (SustainIT 2010)*, USENIX, p. 1-8, 2010.

[CTI 98] CTI Comete (CENT, LIRMM, TIMA, IRESTE, IRISA, LAMI), *CODESIGN, Conception conjointe logiciel-matériel*, Eyrolles, Paris, June 1998.

[DEJ 07] Dejonghe A., Bougard B., Pollin S., Craninckx J., Bourdoux A., Van der Perre L., Catthoor F., "Green reconfigurable Radio Systems. Creating and managing flexibility to overcome battery and spectrum scarcity", *IEEE Signal Processing Magazine, IEEE Society*, vol. 3, p. 90-101, May 2007.

[DEL 07] Delahaye J.P., Plate-forme hétérogène reconfigurable : application à la radio logicielle, Doctoral thesis, University of Rennes 1, April 2007.

[DHI 10] DHINGRA R., NAIDU S., UPRETI G., SAWHNEY R., "Sustainable nanotechnology: trough green methods and life-cycle thinking", *Sustainability*, MDPI, vol. 2, no. 10, p. 3323-3338, 2010.

[DUQ 10] DUQUE CICERI N., GUTOWSKI T.G., GARETTI M., "A tool to estimate materials and manufacturing energy for a product", *Proceedings of the International Symposium on Sustainable Systems and Technology (ISSST 2010), IEEE Computer Society*, p. 1-6, 2010.

[EUG 08] EUGSTER M., HUABO D., JINHUI L., PERERA O., POTTS J., YANG W., Sustainable electronics and electrical equipment for China and the World. A commodity chain sustainability analysis of key Chinese EEE product chains, Report of the International institute for Sustainable Development, 2008.

[FLI 08] FLIPO F., GOSSART C., "Infrastructure numérique et environnement : l'impossible domestication de l'effet rebond", *Actes du Colloque international Services, innovation et développement durable*, 2008.

[FLI 09] FLIPO F., GOSSART C., DELTOUR F., GOURVENNEC B., DOBRÉ M., MICHOT M., BERTHET L., Technologies numériques et crise environnementale : peut-on croire aux TIC vertes ?, Final report of the project Ecotic, Institut Telecom, 2009.

[HER 07] HEART S., "Sustainable Management of Electronic Waste (E-Waste)", *Clean*, Wiley InterScience, vol. 35, no. 4, p. 305-310, 2007.

[HUA 08] HUANG E., TRUONG K., "Breaking the disposable technology paradigm: opportunities for sustainable interaction design for mobile phones", *Proceeding of the Twenty-Sixth Annual SIGCHI Conference on Human Factors in Computing Systems (CHI 2008)*, ACM, p. 323-332, 2008.

[HUA 09] HUANG K., GUO J., XU Z., "Recycling of waste printed circuit boards: A review of current technologies and treatment status in China", *Journal of Hazardous Materials*, vol. 164, p. 3999-408, Elsevier, Paris, 2009.

[JOL 10] JOLLIET O., SAADRÉ M., CRETTAZ P., SHAKED S., *Analyse du cycle de vie. Comprendre et réaliser un écobilan*, 2nd edition, updated and supplemented, Presses polytechniques et universitaires romandes, Lausanne, 2010.

[KEM 05] KEMMA R., VAN ELBURG M., LI W., VAN HOLSTEIJN R., Methodology study eco-design of energy-using products. MEEuP methodology report, Van Holsteijn and Kemma BV, Netherlands, 2005.

[KRI 08] KRISHNAN N., BOYD S., SOMANI A., RAOUX S., CLARK D., DORNFELD D., "A hybrid life cycle inventory of nano-scale semiconductor manufacturing", *Enviro. Sci. and Technol*, vol. 42, p. 3069-3075, 2008.

[KUE 03] KUEHR R., WILLIAMS E., "Computers and the environment: understanding and managing their impact", *Eco-Efficiency in Industry and Science Series*, vol. 14, Kluwer Academic Publishers, Dordrecht, 2003.

[LAS 10] LASETER T., OVCHINNIKOV A., RAZ G., "Reduce, reuse, recycle or rethink", *Strategy + Business*, vol. 61, 2010.

[LEH 10] LEHMAN T., HAMILTON T.J., "Integrated circuits towards reducing E-waste: future design directions", *Proceedings of International Conference on Green Circuits and Systems (ICGCS 2010)*, p. 469-472, 2010.

[LI 10a] LI X., ORTIZ P.J., BROWNE J., FRANKLIN D., OLIVER J.Y., "Smartphone Evolution and reuse: establishing a more sustainable model", *Proceedings of the 39th IEEE International Conference on Parallel Processing Workshop (ICPPW 2010)*, *IEEE Computer Society*, p. 476-484, 2010.

[LI 10b] LI X., ORTIZ P.J., BROWNE J., FRANKLIN D., OLIVER J.Y., "A case for smartphone reuse to augment elementary school education", *Proceedings of the International Conference on Green Computing (GREENCOMP 2010)*, *IEEE Computer Society*, p. 459-466, 2010.

[MAG 08] MAGHREBI H., Etude et implantation FPGA d'un algorithme de routage auto-adaptatif pour réseau de capteurs sans fil, Masters Thesis, SUP'COM, Tunis, June 2008.

[MAX 04] MAXFIELD C., *The Design Warrior's Guide to FPGAs*, Elsevier, Paris, 2004.

[MCC 93] Mcc, Environmental Consciouness; A Strategic Competitiveness Issue for the Electronics and Computer Industry, Microelectronics and Computer Technology corporation (MCC) Report, 1993.

[MUR 01] MURPHY C.F., Electronics, in International Research Institute, World Technlogy (WTEC) Division, WTEC Panel Report on: Environmentally Benign Manufacturing (EBM), p. 81-93, 2001.

[MUR 03] MURPHY C.F., KENIG G.A., ALLEN D.T., LAURENT J.P., DYER D.E., "Development of Parametric Materials, Energy and Emission inventories for Wafer Fabrication in the Semiconductor Industry", *Enviro. Sci. and Technol*, vol. 37, p. 5373-5382, 2003.

[NIG 10] NIGGESCHMIDT S., HELU M., DIAZ N., BEHMANN B., LANZA G., DORNFELD D.A., "Integrating green and sustainability aspects into Life Cycle Performance evaluation", *Proceeding 17th CIRP International Conference on Life Cycle Engineering*, p. 366-371, 2010.

[NOK 10] NOKIA, "Creating our products: Environmental impact", http://www.nokia.com/environment/devices-and-services/creat ing-our-products/environmental-impact, NOKIA corporate Website, 2010.

[OLI 07] OLIVER J.Y., AMIRTHARAJAH R., AKELLA V., "Life cycle aware computing: reusing silicon technology", *Computer, IEEE Computer Society*, vol. 40, no. 12, p. 56-61, 2007.

[RIB 09] RIBON A., Etude et implantation FPGA d'un algorithme de routage auto-adaptatif pour réseau de capteurs sans fil, Masters Thesis, University of Bordeaux, Talence, September 2009.

[SAL 08] SALLINGBOE K., "Where does your mobile phone go to die?", *Mobile Enterprise Magazine*, http://www.greenmobile.co .uk/images/news/memarticle.pdf, 2008.

[SCH 05] SCHWARZER S., DE BONO A., PEDUZZI R., GIULIANI G., KLUSER S., "e-Waste, the hidden side of IT equipment's manufacturing and use", *UNEP Early Warning on Emerging Environmental Threats*, no. 5, 2005.

[SHA 04] SHADMAN F., MCMANUS T.J., "Comment on the 1.7 kilogram microchip: energy and material use in the production of semiconductor devices", *Environment Science and Technology*, *ACS*, vol. 38, no. 6, p. 1915, 2004.

[SUN 09] SUNDARAMOORTHY N., Simplifying embedded hardware and software development with targeted reference designs, Xilinx White Paper, December 2009.

[ULM 04] ULMANN M., HÜBNER M., GRIMM B., BECKER J., "An FPGA Run-Time System for Dynamical On-Demand Reconfiguration", *Proceedings of the 18th International Parallel and Distributed Processing Symposium (IPDPS 2004)*, p. 135-143, 2004.

[WAN 10] WANG W., TEH W.H., "Green energy harvesting technology in 3D IC", *Proceedings of International Conference on Green Circuits and Systems (ICGCS)*, p. 5-8, 2010.

[WIL 02] WILLIAMS E.D., AYRES R.U., HELLER M., "The 1.7 kilogram microchip: energy and material use in the production of semiconductor devices", *Environment Science and Technology*, *ACS*, vol. 36, no. 24, p. 5504-5510, 2002.

[WIL 04] WILLIAMS E.D., AYRES R.U., HELLER M., "Response to comment on the 1.7 kilogram microchip: energy and material use in the production of semiconductor devices", *Environment Science and Technology*, *ACS*, vol. 38, no. 6, p. 1916-1917, 2004.

[XIL 10] XILINX COPORATION, http://www.xilinx.com, 2010.

[ZAD 10] ZADOK G., PUSTINEN R., "The green switch: designing for sustainability in mobile computing", *Proceedings of the first USENIX Sustainable IT Workshop (SustainIT 2010)*, USENIX Association, p. 1-8, 2010.

PART 3

Research Projects on Green Networking Conducted by Industrial Actors

Chapter 8

Schemes for Putting Base Stations in Sleep Mode in Mobile Networks: Presentation and Evaluation

8.1. Motivation

With the worldwide success of smartphones, mobile operators have seen a considerable increase in the data traffic transported on their mobile networks. According to a study by CISCO [CIS 12] on this topic, mobile data traffic tripled in 2010 and is expected to multiply 26-fold between 2010 and 2015. This raises the question of saturation of the networks – therefore, operators have begun looking for solutions to update the mobile network. The solutions proposed to increase the capacity of the network range from adding sectors and/or frequencies to the installation of an additional network layer with LTE-A technology.

In addition, the energy consumption of telecommunications networks has aroused a growing interest over recent years. Environmental protection and

---

Chapter written by Louai SAKER, Salah Eddine ELAYOUBI and Tijani CHAHED.

climate change have become issues of worldwide interest, and a major issue in industrial strategies, including those for ICT, which is responsible for over 2% of CO_2 emissions worldwide [SCH 09]. With the proliferation of access technologies, the amount of energy needed to deploy, maintain and operate these networks is continuously increasing. Reducing the energy consumption of the mobile network has become one of the primary objectives for mobile operators and suppliers of telecoms equipment.

Mobile operators have begun to include environmental responsibility in their strategies, vowing to reduce their CO_2 emissions. This chapter discusses one of the flagship solutions, which can drastically reduce energy consumption: putting networking equipment in sleep mode – particularly base transceiver stations (BTSs).

8.2. Putting macro base transceiver stations in sleep mode

8.2.1. *Structure of the base transceiver station*

In a mobile network, the BTSs are the most energy-hungry nodes (representing nearly 80% of total consumption) [SAK 10]. The architecture of a macro BTS is shown in Figure 8.1. It is divided into two main parts:

– the radio module: this part is essentially made up of a digital signal processing module (information processing and encoding), transport, TRXs (transmission/reception of the signal) and a power amplifier;

– support system: this part essentially comprises the monitoring and control module, the cooling system, the backup battery, rectifiers and other auxiliary equipment. This part is shared between all sectors of the BTS.

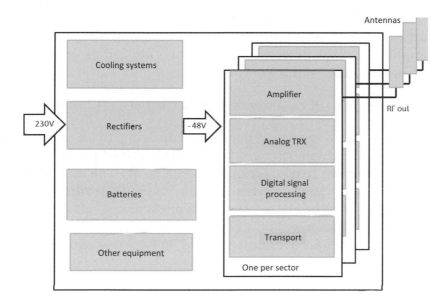

Figure 8.1. *Diagram of a base transceiver station
(three sectors with one amplifier per sector)*

8.2.2. *Model of energy consumption of the BTS*

The model of the consumption of the BTS used in this study is based on measurements. In this model, the energy consumption of the emergency battery and the cooling system is not taken into account, because the extent of this consumption depends largely on the environment. The model essentially takes account of the radio module (amplifier, signal processing system, TRX, etc.). In this consumption, the power amplifiers are the greatest consumers of energy; for instance, for a 3G BTS, the amplifier accounts for between 50 and 65% of the total consumption of the BTS [ABI 10].

The model of the BTS's energy consumption is divided into two components: the first, P_{load}, is the energy consumption which depends on the workload (K), and which consequently varies depending on the traffic. The second part, P_0, is the consumption used to supply the equipment (amplifier, signal processing, transport, etc.); it is constant and independent of the workload:

$$P_{BTS} = P_0 + K\,P_{load} \text{ where } K \text{ varies between 0.1 and 1}$$

EXAMPLE.– Consider a node B comprising three sectors, each containing two carriers with one amplifier per carrier, whose maximum transmission power is 20 W. The minimum energy consumption, corresponding to null traffic (load $K = 0.1$ because the shared channels are still active), is equal to 744.28 W. The maximum energy consumption (load $K = 1$) is equal to 1,052.85 W (a difference of 30%).

8.2.3. *Principle of putting BTSs in sleep mode*

The variation in traffic depends mainly on the time of day. For example, at night (between midnight and 07:00), we see a sharp drop in traffic on the network. However, the energy consumption remains high, due to the fact that the BTS requires a constant power supply to ensure its equipment works (the amplifier, the transport network, etc.) even when there is no traffic. During slack times, certain resources in the mobile network are not used to provide service to users, but nevertheless continually consume energy. Ideally, the evolution of the energy consumption should trace that of the traffic. The top curve in Figure 8.2 represents energy consumption throughout the day. However, even during the hours of low traffic, the network's energy consumption remains relatively high. This is caused by the fact that the BTS's energy consumption is dependent not solely on the load, but that rather, part of it is independent of the load, which accounts for over 70% of the BTS's total consumption base, as discussed above.

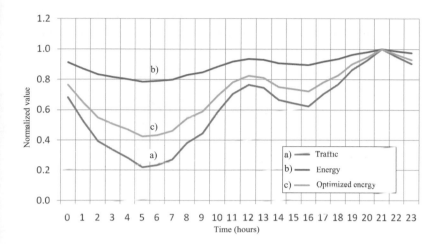

Figure 8.2. *Traffic and energy. The bottom curve represents the average traffic throughout the day for a given site. The middle curve represents the ideal variation in consumption based on the workload, whereas the actual evolution in today's networks is illustrated by the top curve*

In order to greatly reduce the overall consumption of mobile networks whilst limiting the impact on QoS, we apply sleep mode on the scale of the network, turning off certain inactive resources in the network in places or at times of low activity. The resources are the transmitters/receivers (TRX) in the 2G network, the carriers in the 3G/HSDPA network, or even a whole system (2G or 3G when both systems are in place on the same site).

8.2.4. *Illustration of sleep mode. Case of multisystem 2G/3G networks*

As discussed previously, an operator has a set of resources which are active no matter what the volume of traffic passing through, and thus the actual need. In the case of the cooperative 2G/3G network [SAK 09], the BTSs can put one of the two systems (2G or 3G) in sleep mode. In this example we consider a site where two BTSs coexist – one for 2G (x TRX) and the other 3G. The evolution of voice traffic, taken from the counters of the network, is illustrated in Figure 8.3.

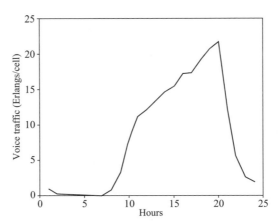

Figure 8.3. *Average voice traffic over the course of one day*

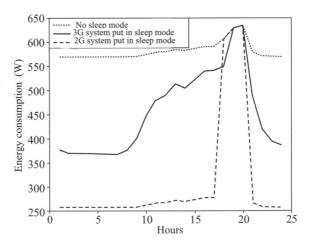

Figure 8.4. *Sleep mode usage in the case of the heterogeneous 2G/3G network*

Figure 8.4 presents the energy consumption of the BTS corresponding to this evolution in traffic for three scenarios: placing the 3G system in sleep mode, placing the 2G system in sleep mode, and the basic scenario (where nothing is put in sleep mode). Placing the 2G system in sleep mode offers a greater gain than doing so for the 3G system, as illustrated in Figure 8.4, because the 2G station in question is more energy-hungry (a power amplifier is used for each TRX with a maximum power of 20 W/TRX). It should be noted that during times of heavy traffic, sleep mode cannot be applied, because all the resources are being used to serve the clients.

8.2.5. *Implementation of sleep mode*

We propose two mechanisms of application for use of sleep mode on a certain number of resources in a given system:

– dynamic sleep mode implementation: the resources are activated/deactivated in real time, depending on the instantaneous workload in the cell. The timescale of activation/deactivation of the resources is a few minutes, depending on the pace of arrivals and departures of calls in the cell;

– semi-static sleep mode implementation: the resources are kept in sleep mode over longer time periods (of around a half hour) in order to minimize the number of commands to activate/deactivate the resources. This mechanism is simpler than that of the dynamic sleep mode implementation.

In what follows, we apply these two proposed methods to the 2G and HSDPA systems.

8.2.5.1. *HSDPA*

Sleep mode implementation in the case of HSDPA consists of activating/deactivating certain carriers (each with a 5 MHz capacity) depending on the traffic. Figure 8.6 shows the energy consumption for the three scenarios corresponding to the average data traffic experienced over a 24-hour period. We can see that the dynamic sleep mode implementation yields a significant gain throughout the day, whereas the semi-static sleep mode implementation is only beneficial when the average traffic is low.

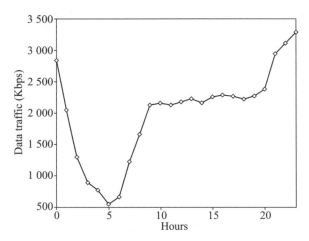

Figure 8.5. *The average data traffic over the course of a day*

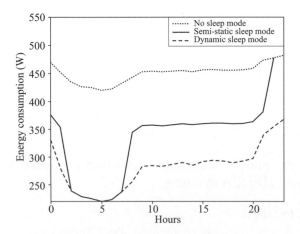

Figure 8.6. *HSDPA energy consumption for the three scenarios: dynamic sleep mode; semi-static sleep mode; no sleep mode*

8.2.5.2. *GSM*

For 2G, we consider a BTS with four TRXs (each operating on a 200 KHz carrier). Figure 8.7 presents the energy consumption of the 2G base transceiver station corresponding to the average traffic for the vocal service for a day (see Figure 8.3). The energy consumption is greatly reduced with the application of sleep mode. During the times when traffic is low (from 21:00 to 05:00), the two mechanisms proposed above yield the same gain, whereas during peak times, dynamic sleep mode offers a greater gain than semi-static sleep mode. It should be noted that semi-static mode offers a gain during peak times, which suggests that some of the BTS's resources are never used (the station is overdimensioned).

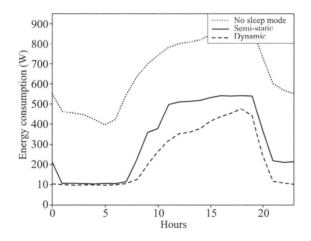

Figure 8.7. *Energy consumption in 2G for the three scenarios: dynamic sleep mode; semi-static sleep mode; no sleep mode*

8.3. Sleep mode in small-cell heterogeneous networks

In the previous section, we applied sleep mode to base transceiver stations in the classic macrocell network. Yet the current architecture of the mobile network cannot sustain the exponential growth in mobile traffic which is sure to continue for the foreseeable near future. Thus it is necessary to improve the capacity on the current mobile network in order to support the increase in traffic. In the futuristic technology of LTE-Advanced, the deployment of small cells constitutes an effective solution to resolve the problem of saturation of the mobile network's capacity. These small cells also enable us to extend the coverage of cellular networks if they are installed on the edge of the macrocell's range.

Picocells are the main types of small cells, and are usually deployed outside or inside certain public buildings (offices, shopping malls, rail stations, airports, etc.). Their range is around 100 m, and their cost and transmission power is low. Picos are installed by the mobile operator, usually in urban areas to alleviate part of the traffic from the macro BTSs.

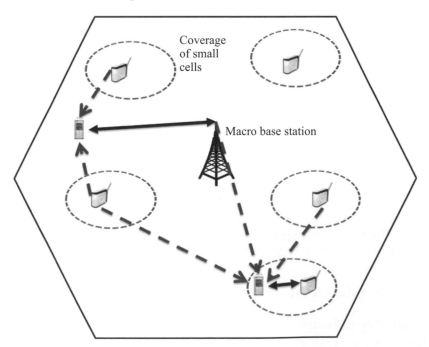

Figure 8.8. *A heterogeneous network made up of macro stations and small cells*

8.3.1. *Energy efficiency of small cells*

Over the course of the past few years, there has been a growing interest in the impact of deploying small cells on the energy efficiency and the performance of the mobile network. Recent studies [RIC 09; SAK 11a; SAK 11b] show that the new cellular architecture based on the dense deployment of small cells is relatively energy efficient in comparison with the current cellular architecture. The energy efficiency of small cells in heterogeneous networks is improved in relation to a solely macrocellular network, as long as the energy consumption of the small cells remains low.

Before going on to look at the schemes for putting small cells in sleep mode, in this section we quantify the impact of the use of small cells on the energy consumption and the performance of the mobile network. The comparison criterion which we shall use, therefore, is "energy efficiency", which gives the traffic transported over the network per unit power. The coverage is calculated using a static simulator (link budget), while the capacity is calculated using queuing theory, where each cell (pico or macro) is modeled as a Processor Sharing (PS) queue (Figure 8.9).

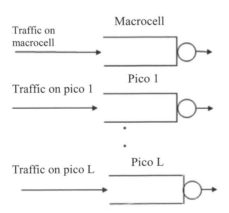

Figure 8.9. *Model of the system (macro with L picos) as a set of queues*

In particular, here, we consider three different densities of picos:

– *light-picos*: two picos per macro sector;

– *medium-picos*: four picos per macro sector;

– *high-picos*: ten picos per macro sector.

The energy efficiency is defined as being the amount of traffic which can be served per Watt. Table 8.1 presents the results in performance and energy efficiency for the deployment of a certain number of picos for each macro sector. Given that they increase the total capacity of the network with a relatively low power consumption per pico, picos constitute an energy-efficient solution because the energy efficiency increases with their deployment.

Energy efficiency (Kbps/W)	Capacity (Mbps)	10W	15W	20W	30W	40W
Zero picos	9.2	12.45	12.45	12.45	12.45	12.45
2 picos	10.3	13.47	13.3	13.13	12.8	12.49
4 picos	11.3	14.31	13.96	13.62	12.99	12.42
10 picos	11.4	13.42	12.42	12.01	10.86	9.91

Table 8.1. *Capacity and energy efficiency for the deployment of picos. The capacity is defined as the maximum amount of traffic served by the network such that the data rate is > 750 Kbps for at least 95% of users. The network under consideration is LTE with a bandwidth between 20 MHz and 2.6 GHz*

The performance results cited in Table 8.1 show that the capacity increases as a function of the number of picos deployed in the cell. However, this increase slows down when the number of picos becomes very great (it becomes almost stable between four and ten picos per macro sector). This is due to the fact that the increased number of picos leads to a geographic reuse of the spectrum (the picos reuse the spectrum of the macros in order to serve their clients), but also generates more interference. When the network becomes very dense, the beneficial effect of the reuse of the spectrum is counterbalanced by the negative effect of the increase in the level of interference.

As regards the energy efficiency, it varies depending on two parameters: the density of picos and their consumption per unit. In terms of the density of picos, the capacity becomes stable for a large number of picos, as previously indicated, and because the addition of each small cell entails an additional consumption in energy, the efficiency increases to begin with, and then decreases when the network becomes too dense. The effect of the unitary consumption of the picos is easier to explain: the less energy the small cell consumes, the more energy-efficient the network is.

8.3.2. *Putting small cells in sleep mode*

The above result shows the energy efficiency of the network when the traffic is high (at peak times when the network approaches its maximum capacity). During slack times, when the traffic is low, the small cells are not needed to carry the traffic but continue to consume energy. Hence, in order to improve the energy efficiency, we introduce sleep mode for the small cells.

In order to study the impact of introducing sleep mode for picos, we consider the average data traffic experienced during the course of a day (Figure 8.5). Figure 8.10 represents the number of picos to be activated throughout the day in order to attain the target QoS. We note that the maximum number of picos required during peak times is nine, whereas at times of low traffic, only three or six are needed, or indeed none at all. Figure 8.11 shows the energy consumption of the cell (picos + macro) with and without sleep mode for the picos. The consumption decreases and our calculation of the average energy efficiency shows a 15% increase.

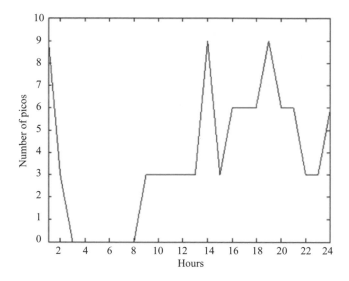

Figure 8.10. *Number of active picos needed*

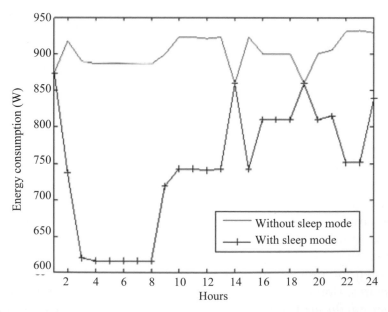

Figure 8.11. *Energy consumption with and without putting the picos in sleep mode*

8.4. Conclusion and considerations on implementation

In this chapter, we have presented the mechanisms of sleep mode implementation to save energy in the mobile network, whilst guaranteeing that the users receive an acceptable Quality of Service. We have applied this scheme to a conventional macrocellular network and a heterogeneous network comprising a macro BTS and picocells. Putting macro-BTSs in sleep modes provides a gain of around 30-40% depending on the variation in traffic. In the context of LTE-Advanced, we have shown that the deployment of small cells increases the capacity of the network but is not always an energy-efficient solution. Thus, sleep mode implementation for these small cells would enable us to increase capacity while preserving energy-efficiency.

In closing, let us point out that the schemes for sleep mode discussed in this chapter run into different practical difficulties in their implementation. For instance, we have shown that, in a multisystem BTS (2G/3G), putting the 2G system in sleep mode offered considerable gains; yet this can only be done once all mobiles have been renewed so that they are able to connect to the 3G system. Another example is that of HSDPA, where we assumed a modular architecture wherein certain carriers can be placed in sleep mode, which involves having one power amplifier per carrier. However, nowadays, constructors are tending to build multicarrier amplifiers, covering up to three carriers on the same frequency band and rendering sleep mode an impossibility on the new BTSs. This does not mean that the mechanisms for sleep mode implementation will not be a reality in the future, but rather that they must be adapted locally to the configuration of the network. An example of a promising scenario is where there are several 3G layers (900 MHz and 2100 MHz, for instance), where it is essential to deploy at least two amplifiers – one per band. Another example is that of LTE, which is currently being rolled out in networks, where each station has several transmission channels in order to be able to exploit the advantages of MIMO (Multiple Input Multiple Output). Adapted to the case of LTE, sleep mode implementation then consists of deactivating certain transmission channels when the traffic is low.

8.5. Bibliography

[ABI 10] ABI RESEARCH, Equipment and RF Power Device Analysis for Cellular and Mobile Wireless Infrastructure Markets, 4Q 2010.

[CIS 12] CISCO, Visual Network Index: Global Mobile Data Traffic Forecast Update, 2011-2016, White Paper, February 2012.

[RIC 09] RICHTER F., FEHSKE A.J., FETTWEIS G.P., "Energy efficiency aspects of base station deployment strategies in cellular networks", *Proceedings of the 70th Vehicular Technology Conference (VTC Fall)*, September 2009.

[SAK 09] SAKER L., ELAYOUBI S., SCHECK H.O., "System selection and sleep mode for energy saving in cooperative 2G/3G networks", *IEEE VTC*, Anchorage, September, 2009.

[SAK 10] SAKER L., ELAYOUBI S.E., "Sleep mode implementation issues in green base stations", *IEEE PIMRC*, Istanbul, September 2010.

[SAK 11a] SAKER L., ELAYOUBI S.E., RONG L., CHAHED T., "Capacity and energy efficiency of picocell deployment in LTE-A networks", *IEEE VTC*, Budapest, May 2011.

[SAK 11b] SAKER L., ELAYOUBI S.E., CHAHED T., "How femtocells impact the capacity and the energy efficiency of LTE-Advanced networks", *IEEE PIMRC*, Toronto, Canada, September 2011.

[SCH 09] SCHECK H.O., "CO_2 Footprint of Cellular Networks", *The Green Base Station Conference*, Bath, UK, April 2009.

Chapter 9

Industrial Application of Green Networking: Smarter Cities

9.1. Introduction

Green networking covers a set of techniques and networking protocols intended to minimize the energy consumption of systems and equipments. The green approach involves both altering behavior and usage, and technological and financial investment. For individuals and for institutions (e.g. local collectives, large groups), adopting green technology can help to reduce their carbon footprint and their energy expenditure. For the new generation of infrastructures – particularly for "smart cities" – green networking represents a springboard so that all machines that are deployed and form systems can provide energy-efficient applications which improve the comfort and safety of the citizens.

Chapter written by Vincent GAY, Paolo MEDAGLIANI, Florian BROEKAERT, Jérémie LEGUAY and Mario LOPEZ RAMOS.

To begin with, this chapter presents an overview of the use of green networking in the framework of the infrastructures of smart cities. Secondly, we review contributions on green networking in this context: low-consumption communication protocols, assisted deployment of a sensor network, low-consumption processor treatments, and finally the integration and use of sensors to aid in decision-making on energy efficiency policies.

9.2. Smart cities and green networking

The infrastructure of a city denotes the set of resources and networks used to provide services to the economic actors and local users (e.g. transport operators, information services, means of security, etc.). With the development of ubiquitously connected information technology and mobile terminals, a new generation of urban infrastructures is coming to life, involving new uses and a wider variety of actors.

The smart city brings a multitude of resources into play – potentially very diverse – such as:

– a set of cameras which can provide video feeds – for instance, to monitor traffic conditions at the "neural points" of the road network or to see emergency situations;

– fixed sensors, which feed back situational data: for example, they might indicate how many parking spaces are available, weather conditions, the level of pollution, the light levels (useful for street lighting), the breakout of a fire, or CBRN (chemical, biological, radiological and nuclear) risks for emergencies;

– public transport (e.g. the bus or metro network), self-service bicycle or electric car hire systems, staff provided with professional applications on their smartphones.

Such resources can be used by multiple actors – particularly:

– organizations which deal with crisis operations (e.g. state, civil security, fire brigade, police forces) at all levels of command (strategic, operational or tactical);

– local government agencies to manage services in the city (public transport, traffic lights, weather forecasting, pollution, events, etc.);

– tertiary companies providing services which reuse the city's municipal services (e.g. the infrastructure operators, availability of parking spaces or self-service equipment);

– applications developed by the citizens and which use real-time, freely-available data (we speak of *open data* [OPE]) made available by urban operators (e.g. parking, transport) or institutions (e.g. state, city authorities).

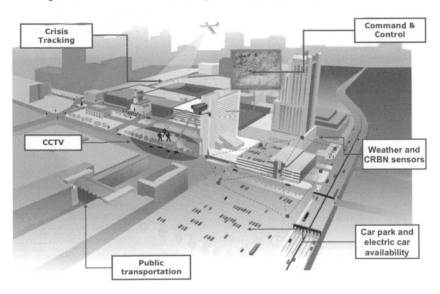

Figure 9.1. *View of the urban infrastructure of the future*

For urban infrastructures, many opportunities arise from the deployment of communicating components, stemming from new uses and economic models. Perspectives are important to construct a new ecosystem involving very diverse actors (e.g. private individuals/users, infrastructure operators, developers of tertiary applications, civil security forces, etc.) which allows us to improve the safety and comfort of the citizens. However, there are considerable technical and organizational challenges to overcome in order to ensure the availability and secure sharing of common resources (for example, sensors, computation facilities, data) used by applications which obey diverse constraints (e.g. protection, reliability, latency, etc.).

Figure 9.2. *"Professional" view of the "smart city" case study*

The development of new generations of urban infrastructure is heavily influenced by green networking. Indeed, the use of resources can and must be reduced by using energy-efficient techniques, whether at the level of processing (e.g. Cloud computing, low-consumption processors), of communication (e.g. MAC protocols, routing in wireless networks), tools for assisting the deployment of sensors/actuators (e.g. optimization of the dimensions and parameters) or by interconnection and system formation (e.g. integration and use with command centers).

9.3. Techniques involved

The concept of a smart city is based on the imagination and implementation of new generations of urban infrastructures. These infrastructures will employ a multitude of technologies and procedures related to green networking. In the next part of this section, we shall detail certain techniques enabling us to optimize the use of resources, i.e. reduce the number or consumption of the pieces of equipment.

9.3.1. *Low-consumption communication protocols*

In the infrastructure of a smart city, an operator will deploy a multitude of sensors in areas of interest. These sensors enable us to collect measurements and data about their environment which enrich the operational view. They may be connected to a sink using low-consumption technologies such as IEEE 802.15.4 or IEEE 802.11 (Low Power Wi-Fi).

In order to extend the radio coverage, these sensors are interlinked to form a network and communicate along multi-hop paths. Most of the time they operate on battery power so as to reduce the amount of cabling in the urban infrastructure, or so as to be able to withstand powercuts. These sensors are to be found, for example, in parking management systems (Hikob, WorldSensing) or surveillance systems (fire alarms, burglar alarms).

For battery-operated equipment, such as Crossbow's MICAz [MIC], TelosB or Open Mote [OPE b], communication represents a large part of the operational energy budget, often accounting for more power than the computations and the sampling of the sensor module. So as to increase the lifetime of a sensor network, a number of energy-saving techniques are possible, to adapt the transmission power, put the radio chip into sleep mode or route messages in view of the remaining energy of the nodes.

9.3.1.1. *Related works*

MAC protocols and routing protocols for wireless sensor networks are very fertile fields of research. The literature in this domain is particularly abundant, as demonstrated by [MAC], which brings together information about the best-known MAC protocols and proposes a classification system for them. In [ITE 11], the ITEA2 Geodes project produced a manual about a vast portfolio of energy-saving techniques – particularly on the scale of a network. This review includes techniques operating at the PHY (hardware) and link levels: MIMO transmission, adaptation of the transmission power, error control mechanism; MAC protocols based on cycles of activity/sleep of the radio chip; and routing mechanisms which take account of the amount of energy remaining in the nodes involved in the transmission and relaying of messages.

9.3.1.2. *Cascade-MAC protocol*

A predominant traffic model in a WSN (Wireless Sensor Network) is to channel data from sources towards a central node called a *sink*. Figure 9.3 shows an example of a tree structure in such a scenario. Supposing the paths are relatively stable, an interesting approach for the Medium Access Control protocol consists of sequentially waking up the nodes which are along a path in the direction of the sink. Thus, a one-hop latency is reduced approximately to the delay introduced between the cycles of activity of the radio, independently of the length of the cycle. The D-MAC protocol describes this approach and evaluates an implementation in the *ns-2* simulator, supposing that the nodes are synchronized in an omniscient manner.

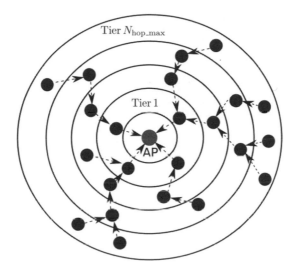

Figure 9.3. *Topology of a sensor network used for data collection*

In [MED], Medagliani *et al.* introduce the Cascade-MAC protocol, which is inspired by D-MAC, but offers significant improvements in terms of the choices of implementation, particularly as regards the signaling which allows synchronization between the nodes and the transmission routine. The main characteristics of Cascade-MAC, illustrated in Figure 9.4, are as follows:

– *Staggering of the cycles of radio activity*: each node is aware of the number of hops between it and the sink. The latter periodically begins a phase of explicit synchronization, during which it broadcasts to all the nodes the date of its next wake-up. Thus, each node staggers the moment of its next wake-up with a difference proportional to its depth in the tree;

– *Sending and relaying of a packet*: during each cycle of radio activity, *Send Alarm* is triggered just before the next node, along the path toward the sink, wakes up to monitor activity on the radio alarm (*Polling Alarm*). Thus, the node

with a packet to send dispatches a short preamble so that
the receiving node remains active and can receive the packet.

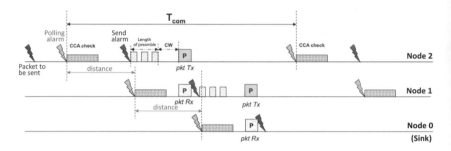

Figure 9.4. *Diagram of cycles of radio activity*
for the Cascade-MAC protocol

The latency of the transmission on the first hop is
expressed thus:

$$D_{\text{Cas-MAC}_{\text{1st hop}}} = \frac{t_{\text{comm}}}{2} + CW + WT_{\text{length}} + S_{\text{d}}$$

where *tcomm / 2* denotes the average time that the
transmitting node has to wait between the moment the
packet to be sent is received from the application (and
buffered) and the triggering of the *Send alarm*; *CW* is the
overall duration of the contention phase (contention
window); *WTlength* is the length of the preamble; and *Sd* is
the sending duration of the packet.

Medagliani *et al.* implemented Cascade-MAC in TinyOS
and conducted measurements of the latency using Crossbow
MICAz nodes and the simulator AvroraZ [AVR]. From their
results, the comparison between Cascade-MAC and the
X-MAC protocol shows that Cascade-MAC guarantees a far
shorter multi-hop latency than X-MAC when the rate of
activity characterizing the radio cycle, denoted *βcomm*, is
less than 10%.

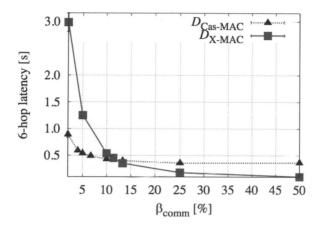

Figure 9.5. *Comparison of the experimental measurements of latency on a 6-hop path between Cascade-MAC (dotted line) and X-MAC (solid line)*

In order for Cascade-MAC to maintain synchronization between the nodes, a scaling down mechanism consists of using the signaling of the routing protocol to set the pace for the cycles of radio activity. Developed by the ROLL group [ROL] within the IETF, the RPL protocol (*Routing Protocol for Low power and lossy networks*) is particularly well adapted to multipoint-to-point data accumulation traffic. The use of DIO packets (*DODAG Information Object*), with the adaptive mechanism of *trickle*, enables a node to select its parent at the routing level – that is, the node to which the packets addressed to the sink are transmitted. By adding information into a DIO packet about the wake-up date, at the MAC level, a node can adjust its cycle to stagger its wake-up in relation to its parent.

Cascade-MAC is an example of a low-consumption radio medium access protocol. Like many MAC protocols, its design is optimized for a certain prevailing type of traffic – here the accumulation of data towards a sink. In the context of a sensor network sending back *in situ* measurements for the purposes of surveillance, it is possible to adjust the behavior of the MAC layer in order to attain predetermined levels of latency.

9.3.2. *Assistance in the deployment of sensor networks*

When installing a sensor network inside an infrastructure, an operator has to decide on the dimensions of the system, the settings and the position of the nodes. These choices will have an impact on the performances of the system, e.g. characterized by the lifetime, the delay to receive an alert, the reliability in terms of detecting an abnormal phenomenon, etc.

Mathematical modeling of the performances, taking account as closely as possible of the operation of the hardware and the environment, is the first step towards assistance in deployment. In conjunction with constrained mathematical optimization functions, this technique can present the operator with the possible configurations of deployment along with the associated QoS. In particular, the operator can see the underlying tradeoffs between lifetime, reactivity, reliability, etc. In addition, the goal for the operator is to determine the optimal configuration in relation to his own needs – that is, to avoid using too many machines or an unsuitable parameter setup.

In the next part of this section, we present the modeling framework developed in the publication [MED], and an iterative procedure of deployment assistance which uses that framework [GAY 11]. The models and procedure of assistance presented below are particularly relevant for surveillance applications (e.g. intrusion detection in an area of the city). However, the advantage to the approach lies in a far wider context.

9.3.2.1. *Mathematical models and optimization*

The analytical framework put forward in [MED] enables us to model a surveillance system in the context of stochastic or deterministic deployment of networked sensors, focusing on the following characteristic performance criteria:

– *Pmd* (*probability of missed detection*): this is the probability that the network will not detect an event (say, the arrival of a target or occurrence of a phenomenon) in the knowledge that it is happening. In particular, this probability depends on the level of vigilance of the nodes (e.g. frequency of sampling of the seismic or infrared signal), on their geographic position and on the capacity of the sensors to work together to correlate their observations;

– *D* (delay of alert transmission): this is the delay in transmission of the alert to the gateway node. This delay usually depends on the MAC (*Medium Access Control*) protocol and on the routing used, as well as the associated parameters;

– *L* (*Lifetime* of the system): this indicator gives the time for which the system is considered to be operational. It may be calculated in a number of different ways – for instance, the time until at least one sensor runs out of energy to operate.

The analytical framework deals with two types of deployment of fixed sensors: stochastic, meaning that the nodes are positioned randomly (e.g. dropped into an area from a helicopter); or deterministic.

Thus, using optimization functions, we can observe the space from optimal configurations based on the objectives and constraints defined by the operator himself. This helps to understand the tradeoffs underlying the system's operation. For instance, the constraints and objectives may be to maximize detection performances in light of constraints relating to delays in the transmission of alerts. Figure 9.6 shows the contour lines of the maximum lifetime (in days) depending on the constraints as regards reliability of detection *(Pmd)* and the delay in transmission *(D)* using X-MAC and Cascade-MAC protocols.

For a given MAC protocol, the operator can see the different points of function which enable a certain value of the maximum lifetime to be reached and, in this particular case, the tradeoff between better reactivity (lower value of D^*) and better reliability (lower value of Pmd^*). Also, between the two MAC protocols, the operator can see that the use of Cascade-MAC allows more stringent constraints to be satisfied than X-MAC, to achieve a maximum lifetime.

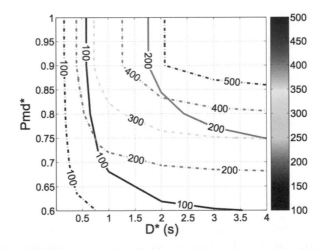

Figure 9.6. *Contour lines of the maximum lifetime (in days) based on the constraints relating to reliability of detection (Pmd) and the delay in transmission (D) (that is, their maximum acceptable values Pmd* and D*), using the X-MAC (solid line) and Cascade-MAC (dotted line) protocols*

In conclusion, modeling the performances of a sensor network in an analytical framework is a difficult task. Although the models for the performance criteria are simplified in some respects, they provide an interesting base from which to evaluate the behavior of a sensor network in a concrete manner for an operator. In particular, the issue is to give a better account of the tradeoffs between the different performance criteria on the one hand, and on the other hand, to evaluate the impact of the choice of a specific protocol.

9.3.2.2. *Iterative configuration process*

The analytical framework presented in section 9.3.2.1 is a first step towards assisting an operator with deployment. However, it is still uncomfortable for an operator who has to make concrete choices as follows: for instance, how many nodes are needed in order to achieve such-and-such a level of reliability? Or, what is the maximum surface area which we can monitor with this level of latency and with this number of nodes at our disposal? In [GAY 11], we present a procedure for configuring a wireless fixed network which is based on the optimization techniques presented in section 9.3.2.1. This procedure helps operators in placing the nodes and configuring the parameters of the system in the framework of particular operational contexts such as area surveillance. In particular, the procedure tells the operator the minimum number of nodes, their position and their parameters in order to attain performance targets across a set of distinct geographical areas. With this procedure, the operator adopts a green approach, in that he avoids using excess machines or configuring them, in doubt, in energy-hungry states which are unnecessary in relation to the operational needs.

The configuration procedure, illustrated by Figure 9.7, is characterized in detail by the following stages:

1) definition of the performance criteria which constitute the constraints, with their associated threshold values, and at least one performance criterion to be optimized, for at least one zone in which nodes are to be installed; each performance criterion is defined by a mathematical model;

2) definition for the zone or for each zone:

a) of the characteristics of the zone,

b) and of the characteristics of the sensors in that zone;

3) allocation of a number of nodes to the zone(s) to be equipped;

4) application of a process of optimization to each zone; the process of optimization by zone involves the following stages:

a) determination of a point of operation if one exists, characterizing the behavior of the sensor network in the area (position of the nodes, parameterization),

b) determination of the configuration of the sensor network in the area if one exists, defining the configuration parameters of the sensor network,

c) determination of the possible need to use more nodes or revise the performance criteria defined in stage 1;

5) increase of the number of nodes or modification of the performance criteria defined in the zone(s) where the performance criteria are not satisfied, and repetition of the process of optimization by zone in these areas, with the new number of nodes or the new performance criteria;

6) application of the configuration thus determined to each node in the zone(s).

Depending on the particular mode of execution, the procedure includes different variants enabling users to determine the minimum number of nodes required or the maximum size of the zone, or even to place a subset of nodes at predetermined positions (e.g. previously-identified points of passage).

For example, to determine the minimum number of nodes required, the optimization by zone process includes a stage of increasing or decreasing the number of nodes to equip the zone, if at least one performance criterion to be optimized is not satisfied. Stages 4a and 4b are then repeated with the newly increased or decreased number of nodes.

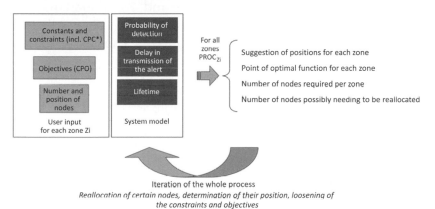

Figure 9.7. *Principle of the procedure of iterative deployment assistance [GAY 11]*

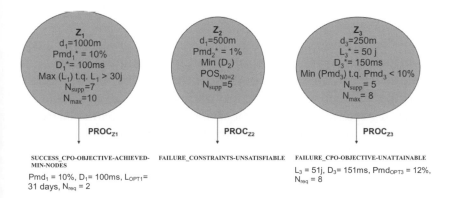

Figure 9.8a. *First iteration of the whole process*

Figures 9.8a and 9.8b show an example of the application of the whole process to three distinct zones – Z_1, Z_2 and Z_3, each with specific characteristics and performance constraints. At the end of the first iteration of the procedure, the operator sees that in zone Z_1, he can afford to reallocate five nodes, and that he has to either use more nodes or loosen his constraints for zones Z_2 and Z_3. Following the second iteration, the operator obtains a realizable configuration for these two zones, having chosen to loosen the constraints on Z_2 and reallocate five nodes to Z_3.

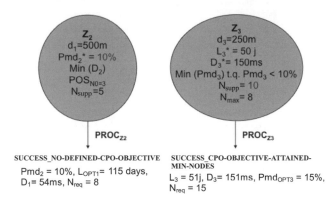

Figure 9.8b. *Second iteration of the whole process*

In conclusion, assisted deployment procedures are essential in order to appropriately configure a set of equipment (in terms of position, number and parameters). The use of such procedures satisfies concrete professional needs. Based on mathematical models for various criteria and proposing optimal configurations, the procedure offers numerous advantages to an infrastructure operator (such as minimization of the system's energy consumption, and therefore an extension of its operational lifetime, use of only the number of machines which are needed, etc.). Such procedures are extensible in that the portfolio of performance criteria can be extended. Furthermore, they can be used before the rollout but also during operation if the operational constraints change.

9.3.3. *Low-consumption processor treatments*

In this section, we shall discuss the use of green networking in the context of low-consumption treatments for embedded systems. Indeed, of the devices which make up the infrastructure of a smart city, it may be interesting to reduce the consumption related to the processor treatment, in video acquisition/transmission or in transceiver stations of PMR (*Professional Mobile Radio*) networks, used by transport operators or the public safety forces. To begin with, we shall describe the more specific context, before going into detail on two techniques in particular.

9.3.3.1. *Context*

Today, the proliferation of electronic systems in our society is causing a constantly-increasing rate of energy consumption. The International Energy Agency predicts global consumption by electronic equipment in 2030 equal to the current domestic consumption of America and Japan. The smart city fits perfectly into this context, where we will find a great many interconnected devices – particularly sensors, actuators, video cameras, BTSs, data servers, PCs for command centers, the smartphones of private users, etc. The power requirement of these devices is growing for the digital part and particularly the processor, as Figure 9.9a shows. In order to cope with this ever-increasing need for new functions and computing power, processor architectures have greatly evolved, currently including multiple cores, graphics processing units (GPUs) and other devoted hardware accelerators.

However, although the new hardware architectures are just about capable of satisfying the performance requirements, the evolution of the technology has not yet produced a good enough technological breakthrough to deal with the current energy demand. Consequently, the autonomy of embedded systems is now under threat.

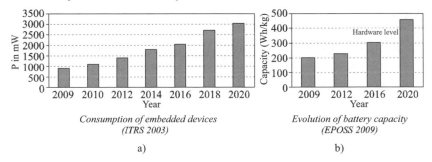

Consumption of embedded devices
(ITRS 2003)

a)

Evolution of battery capacity
(EPOSS 2009)

b)

Figure 9.9. *Increase in devices' consumption in comparison to the evolution in battery capacities*

Thus, the issues of reducing the energy footprint (for systems drawing from the mainline power supply) and increasing autonomy (for battery-powered systems) are beginning to be major ones. In this context, green computing is encouraged by the European Union by way of standards, and by programs such as Energy Star, developed by the Environmental Protection Agency (EPA), to reward high energy-efficiency products. This challenge necessitates techniques to reduce and manage electricity consumption at various levels. In the following section, we shall focus particularly on techniques which apply to the processor.

In a digital circuit, the consumption is attributable to two elements: the *static power* and the *dynamic power*. The evolution of CMOS technology is progressively reducing the size of the transistors. On the one hand, transistors can function at lower voltages, but the disadvantage is that the leakage current (static power) increases exponentially. This static consumption therefore cannot be reduced. However, the dynamic power, for its part, *can* be reduced during the system's run time by using the low-power modes available in most modern hardware architectures, at the level of the processor and on the peripheries. Two approaches can be distinguished:

– dynamic voltage/frequency scaling (DVFS);

– dynamic power management (DPM, or sleep modes, switching off of cores), which consists of cutting the power supply associated with specific areas of the chip.

These two techniques enable us to activate and deactivate components and their operating modes, in order to adapt the system's performance to the actual workload in real time. Indeed, another significant cause of energy inefficiency of systems is the so-called "idle" power, which is wasted when the devices are turned on but are "waiting" idly. Thus, we can use these mechanisms at different levels:

– at the application level, by choosing the level of performance which is best adapted to the needs – e.g. by adapting the modes of the processor in relation to the reception/transmission data rate or to the complexity of the algorithm and the user performances required;

– at the level of the operating system (OS), because information coming from the ensemble of applications and from the use of the hardware components can easily be handled by this central layer of the system.

In standard operating systems (e.g. Windows, Linux), power management regimes are already in existence, but they do not satisfy the requirements of real-time systems because the adaptation of resources is carried out *a posteriori* (that is, heuristics based on past activity). Consequently, future events may be missed if the level of performance is decreased too greatly. The scientific community is therefore active on this subject, and many works have been published about scheduling systems which facilitate management of resources by way of the use of the DVFS and DPM mechanisms, without adversely affecting the performances [ITE 11; BEN 00; GOL 95; RUNTIME; ZAF 05; LU 00; BAM 10].

9.3.3.2. *Dynamic Voltage / Frequency Scaling*

Dynamic Voltage/Frequency Scaling (DVFS) is a technique which enables us to alter the voltage/frequency ratio of the processor during run time, and thus adapt the power of the processor (and therefore its consumption) to the workload actually necessary to execute an application. From an energy point of view, it is indeed more economical to stagger the processes over a long period of time, rather than carry out those processes quickly and then place the processor in an idle state (see Figure 9.10).

Example of a task with a workload W and which must be completed by the date D

Figure 9.10. *The principle of DVFS*

The DVFS mechanism can be used in a number of different ways: statically at the design phase, at the application level or at the level of the operating system.

Statically at the design phase: in this scenario, we determine the minimum frequency that can guarantee the correct execution of the application. We then fix that global frequency. In this way, less energy is used than if the processor were used by default.

At the application level, the application directly makes the request to the DVFS driver. The works taken from [BIL 11] show a usage of DVFS to adapt the processor power to the performances in terms of possible transmission/reception data rate with an implementation in OMAP3530/Linux. The transmission data rate is limited by the bandwidth of the Zigbee. In this particular case, this does not go above 5.6 frames per second. Thus, it is not necessary to constantly use the process at its full capacity (500 MHz) to support this data rate. Thus, we observe that:

1) if the data rate is less than 2 KB/sec, the processor frequency does not need to go above 125 MHz;

2) if the data rate is less than 4 KB/sec, the processor frequency does not need to go above 125 MHz;

3) if the data rate is less than 8 KB/sec, the processor frequency does not need to go above 125 MHz.

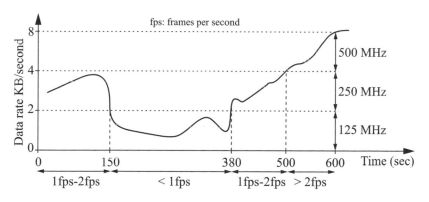

Figure 9.11. *Example of the use of DVFS in relation to the transmission speed*

At the level of the operating system, a framework is in charge of taking decisions for the applications. The framework illustrated in Figure 9.12 was developed with a view to portability on different OS and hardware platforms. It is based on a set of services, enabling it to monitor and reduce the energy consumption of the real-time applications. In particular, the framework is charged with monitoring the workload, the available resources and the performance requirements of the application. In addition, the framework decides when the processors can enter a low-consumption mode. A "low-power" scheduling algorithm using the DVFS (and DPM) mechanisms identifies the time intervals when the resources are not being used ("slack times"). The framework then safely adapts the modes of the processor in accordance with the constraint of the task and the slack period available. For this purpose, it must be aware of the transition times for switching from one mode to another (typically around 100 µs). Also, the use of this framework requires that the application be equipped with "markers" facilitating communication between it and the framework, and finally that the platform be calibrated in order to obtain maximum run-times for the different sections of code of the application.

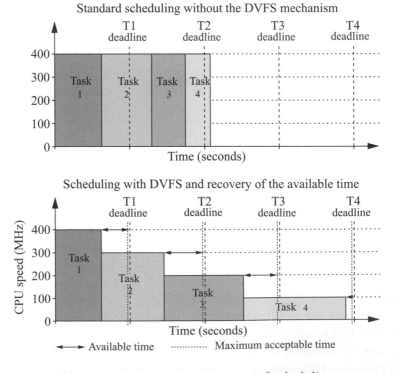

Figure 9.12. *Principle of "low-power" scheduling*

On the applications tested, which are the encoding/ decoding of a video-surveillance feed, we show that the gains in consumption attainable vary between 20% and 40% depending on the QoS requirements of the application. The greater the requirement in terms of QoS, the less time will be spent in the idle state, and the less significant the gain in terms of consumption will be.

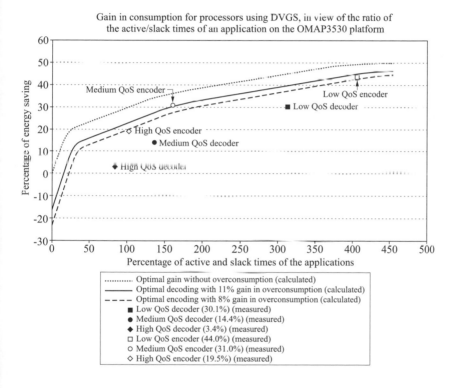

Figure 9.13. *Estimation of the gains in consumption with DVFS on OMAP3530*

More generally, Figure 9.13 enables us to estimate the gains in consumption that it is possible to achieve using the DVFS technique on an OMAP3530 platform based on the following information about the application: *idle time in relation to the activity time*. For this, we operate on the hypothesis that we shall be able to find a point of function which allows us to reduce the idle time to zero. Hence, this gives a preliminary approximation of what can be achieved as a reduction in consumption with the DVFS technique. We can thus see that the limitation comes from the available points of function, and that beyond a ratio of 200%, the gain in consumption no longer changes greatly, stagnating at around 40%.

In order to obtain these gains in consumption, we take measurements of the consumption (or use values taken from the processor documentation) to construct a model of the processor's consumption. This model takes account of the processor's consumption at the different points of operation (voltage/frequency pairs) and in the "0% load" (idle state) and "100% load" modes.

Consumption of the TI OMAP3530 CPU (ARM Cortex™A8)

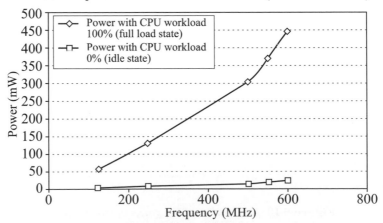

Figure 9.14. *Characteristics of consumption of the ARM Cortex™ A8 processor in the different DVFS modes*

9.3.3.3. *Sleep mode or DPM*

Dynamic Power Management (DPM) is a very similar mechanism to DVFS. There are different sleep states, which are characterized by different levels of consumption and performance. In contrast to DVFS, where the strategy is to reduce the idle time by placing the processor at a frequency which prolongs the duration of the processing to maximum, DPM helps to reduce idle consumption by putting the processor in sleep mode or states of inactivity. The metric used to implement DPM strategies can be summed up as the transition time taken to enter or wake from sleep mode. By way of comparison, the transition time here is around one millisecond, but the consumptions may reach up to 0.2 mW.

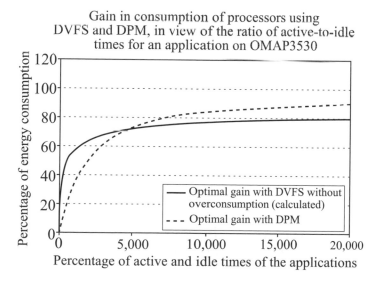

Figure 9.15. *Estimation of the gains in consumption with the DPM mechanism on OMAP3530*

The choice of whether to use DPM or DVFS depends on the constraints of the application. If the application includes long periods of inactivity, choosing DPM over DVFS will offer greater gains in terms of consumption. Conversely, if we wish to preserve the reactivity of the application, we would tend to opt for DVFS. Furthermore, it should be noted that DPM causes an additional issue – that of waking the processor up – which can be done by the use of a timer or triggered by an outside event (alert).

Similarly as for DVFS, Figure 9.15 enables us to approximate the gains in consumption which may be achievable using DPM (on OMAP3530) depending on the ratio between the idle time and the active time. We can see that beyond 5,000%, it is more advantageous to employ DPM than DVFS, and that the gains attainable can be greater than 90%.

In the case of the smart city scenario, the results presented here concerning DVFS and DPM can be used to guide choices as regards the use of one or the other of these mechanisms to reduce the processor consumption of the equipment deployed depending on the application behavior.

9.3.4. *System integration of heterogeneous sensors*

With the recent advances discussed in the previous sections (e.g. low-consumption protocols, miniaturization of hardware), a growing number of embedded devices (such as sensors, actuators, cameras, etc.) will be used in the future infrastructures of cities. In view of the variety of both hardware and software, and of the multitudes of radio technologies (e.g. Bluetooth, WiFi, RFID, PLC, etc.), creating a system integrating a large number of heterogeneous devices is a difficult task. This is a different aspect of green networking to that of designing the architecture of an IT system based on recognized norms.

Given the lifetimes of the machines, the issue is guaranteeing that new technologies and new services will be able to be added into the system over the course of time, without jeopardizing the stability of an infrastructure intended to provide decades of service.

Standard-setting organizations such as IETF or ETSI seek to promote norms at the level of protocols, interfaces or even data models. Based on these norms, the infrastructures of tomorrow will be able to evolve with limited additional investment and modifications: on the one hand, we could have several generations of machines and multiple radio technologies coexisting in the same system, and add new protocols or functions by a simple software update on the other hand. Thus, the adoption of norms is part of the green networking approach, in that it limits wastage caused by the obsolescence of the hardware or incompatibility of the protocols.

Below, we describe the efforts of the IETF concerning IP convergence and Machine-to-Machine (M2M) frameworks facilitating interoperability between heterogeneous equipment.

9.3.4.1. *IP interoperability / continuity*

The IETF (Internet Engineering Task Force) is a group working on elaborating so-called "Internet standards". Faced with the multiplication and variety of embedded devices in so-called *lossy and low-power* networks, the IETF has set up working groups to come up with standards guaranteeing scaling, security and reliability of communications between, to, or from such devices. An excellent idea is to use IP as a reference layer on which to develop protocols for routing, transport or application. IPv6 offers considerable advantages: an address space facilitating the identification of millions of devices; end-to-end support for bidirectional communications from a machine to an Internet-connected

PC; most lightweight embedded operating systems already offer a protocol stack based on IPv6 (e.g. Contiki, TinyOS), etc.

Figure 9.16 shows a protocol stack for embedded equipment which is based on IPv6. Above the PHY/multiple link layers (e.g. Ethernet, WiFi, cellular, WiMAX, CPL, 802.15.4), an adaptation layer for transporting IPv6 packets is given by the existing norms (IETF RFC 2464, 5072, 5121) and in the context of wireless networks IEEE 802.15.4 by the IETF norm 6LoWPAN. At level 3, RPL is a routing protocol defined by the IETF working group called RoLL, for *Routing over Low-power and Lossy networks*.

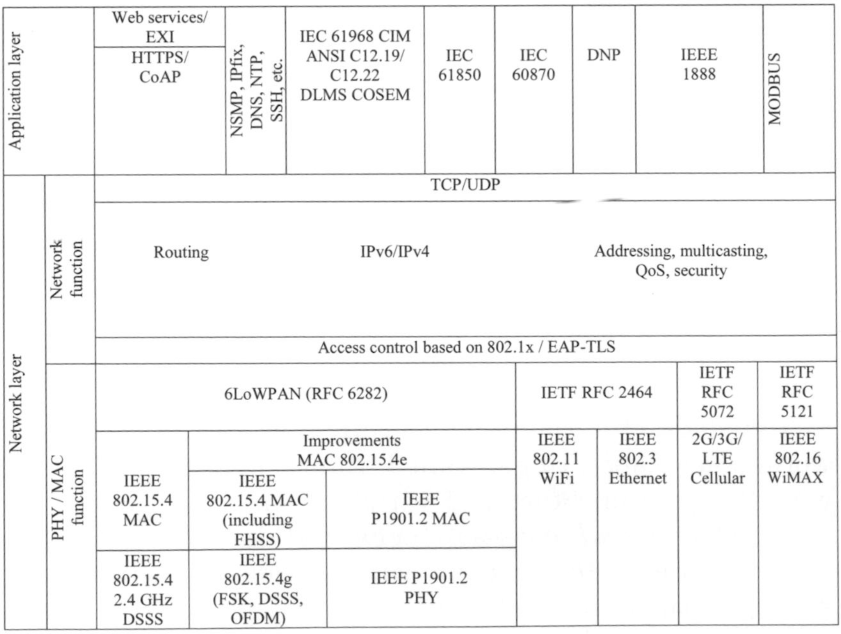

Application layer		Web services/ EXI	NSMP, IPfix, DNS, NTP, SSH, etc.	IEC 61968 CIM	IEC 61850	IEC 60870	DNP	IEEE 1888	MODBUS
		HTTPS/ CoAP		ANSI C12.19/ C12.22 DLMS COSEM					
Network layer	Network function	TCP/UDP							
		Routing		IPv6/IPv4			Addressing, multicasting, QoS, security		
		Access control based on 802.1x / EAP-TLS							
	PHY / MAC function	6LoWPAN (RFC 6282)			IETF RFC 2464			IETF RFC 5072	IETF RFC 5121
		IEEE 802.15.4 MAC	IEEE 802.15.4 MAC (including FHSS)	Improvements MAC 802.15.4e	IEEE P1901.2 MAC	IEEE 802.11 WiFi	IEEE 802.3 Ethernet	2G/3G/ LTE Cellular	IEEE 802.16 WiMAX
		IEEE 802.15.4 2.4 GHz DSSS	IEEE 802.15.4g (FSK, DSSS, OFDM)	IEEE P1901.2 PHY					

Figure 9.16. *Protocol stack used for sensor networks based on IPv6 (source Cisco) [KOP 11]*

6LoWPAN: The IETF working group called 6LoWPAN specifies how IP packets are encapsulated in IEEE 802.15.4.

The goal is to optimize the transmission of IPv6 packets across low-power and lossy networks, like 802.15.4. The working group has published many RFCs (requests for comments), relating particularly to the compression of headers (RFC 6282) which reduces the size of IPv6 headers to 40 bytes and that of UDP headers to 8 bytes; the fragmentation and reassembly of the IPv6 packets; and other functions such as 6LowPAN's detection of duplicate addresses on the diffusion link layers.

Although these mechanisms were initially developed for the norm IEEE 802.15.4, they can be reused by other linking layers which respect the principles of addressing and of the MAC layer of IEEE 802.15.4. Such is the case, for instance, with the CPL linking layer in the norm IEEE P1901.2, the DECT Ultra Low Energy (ULE) standards or indeed Bluetooth Low Energy (LE).

RPL: RPL is a highly flexible and dynamic routing protocol. It was originally designed to cause very little control traffic whilst operating in so-called "difficult" environments, characterized by a low connection speed and potentially elevated error rates. RPL offers numerous advanced functionalities, such as adaptive timers limiting the traffic load on the control plane, support of multiple topologies, loop detection and temporary instability management (by way of global or local repair modes). On the other hand, numerous works are underway – notably in the European project CALIPSO [CAL] – to optimize RPL's function on *duty-cycle* MAC layers (following wakeup and sleep phases).

CoAP: the IETF working group CoRE (*Constrained RESTful Environments*) defined the CoAP standard – *Constrained Application Protocol* – with the aim of supporting REST-type applications in constrained environments identical to those dealt with by the working groups RoLL and 6LowPAN. CoAP is a Web protocol similar to HTTP, to use the functionalities of a constrained device (seen as *Web resources*). In addition, CoAP defines the support of a large-scale resource discovery mechanism; of multicast communications; and of asynchronous exchanges such as *publish / subscribe*.

In conclusion, the IETF currently defines a suite of protocols called "IP", based notably on 6LowPAN, RPL and CoAP, to enable the integration of constrained equipments into an interconnected IT system and facilitate the interoperability of this equipment with Internet applications and machines.

9.3.4.2. *Internet-of-Things and Machine-to-Machine Framework*

The *Internet of Things (IoT)* is a recent concept based on the integration into the Internet of a multitude of communicating objects or "things" (e.g. sensors, actuators, RFID, etc.). Used mainly in the telecoms industry, the term *Machine-to-Machine (M2M)* denotes embedded systems which communicate with other machines (e.g. databases, application servers, smartphones) without human intervention. Beyond the semantic distinctions, what the concepts of *IoT* and *M2M* have in common is that they conceive the architecture of IT systems involving interconnection and interoperability between a multitude of devices, possibly constrained and highly heterogeneous.

By providing standards for the representation of the devices and their data on the one hand, and for interfaces on the other, the immense benefit of *IoT* or *M2M* systems is that they enable engineers to easily integrate new devices, and third parties to develop applications or solutions using this kind of equipment and to get around the hardware and software specificities of that equipment. Owing to the ever-increasing penetration of these machines, the possibilities, both in terms of finance and of usage, are considerable. In green networking circles, the issue is to favor the emergence and adoption of reference standards so that *IoT*- or *M2M*-type infrastructures can easily evolve in decades to come, thereby avoiding costly hardware updates.

Among the standardization efforts concerning *M2M*, we can cite the activity of the European Telecommunications Standards Institute (ETSI), which includes a working group devoted to the topic. In late 2011, ETSI presented the first version (*Release 1.0*) of the ETSI M2M architecture [M2M]. Based on the REST architecture, ETSI M2M defines standard interfaces between equipment, the gateways and the "core" network. In addition, ETSI M2M R1 defines a model to represent the entities (i.e. the equipment, gateways, applications), the data, and the support for a publish/subscribe type notification mechanism. ETSI M2M is based on the HTTP and CoAP protocols for communication between M2M entities.

Cooperation between the different standardization organizations is important in order to facilitate the blooming of the infrastructures of tomorrow. Thus, ETSI M2M attempts to coordinate with the 3GPP, the ZigBee Alliance, the BroadBand Forum (BBF) or the Open Mobile Alliance (OMA) to build bridges between the different standards, and enable a mobile telephone, an ADSL router or a ZigBee machine to be represented with a single data model within the same M2M framework.

9.3.4.3. *Application to energy efficiency of buildings*

The residential and tertiary sector represents over 40% of final energy consumption in the EU. This sector is expanding – a phenomenon which will inevitably lead to an increase in this proportion and in the emissions associated with it. In France, Réglementation Thermique 2012 (RT2012 – Heat Regulation 2012) aims to limit the energy consumption of new buildings: all new constructions must exhibit an average primary energy consumption (that is, before transformation and transport) less than 50 kWh/m^2/year (in the knowledge that today, the average consumption of old buildings is between 200 and 250 kWh/m^2/year). The regulation imposes particular construction techniques (thermal bridges, etc.) the production of renewable energy and mechanisms for measuring or estimating energy consumption per use for the occupant's information.

Yet many problems still remain to be solved: although the law stipulates it, there are as yet no architects or insurers who are able to make a commitment in terms of the long-term energy efficiency of the building constructed. This is due to the fact that there is very little human expertise in the management of low-consumption buildings (design, maintenance, usage) and there are very few technical solutions to closely supervise the energy consumption of the building over the course of time. Such systems would inform the users of their impact on the daily consumption so as to help them reduce it, detect anomalies which might arise as the building ages, and adapt in view not only of the information sent back by sensors, but also from external information sources (such as weather forecasts) and above all the expectations of the occupants.

The main sources of consumption in buildings are: heating and air conditioning, lighting, electrodomestic and electronic equipment, elevators, etc. However, few of these technologies are capable of communication today, and those which are function only in isolated silos, with the providers and standards of communications being heterogeneous.

In the same way as the Ethernet and TCP/IP standards have made terminal networking a commonplace phenomenon, the adoption of 6LoWPAN (be it on IEEE 802.15.4 or CPL), RPL and CoAP could greatly simplify the networking of communication objects for energy management in buildings. However, in the interests of a genuine democratization of these technologies, it is important to reduce the cost of the true added value of the system, the power management software, just as Android facilitated a reduction in the price of smartphones.

The OSAmI project [OSA] aims to provide a modular open-source solution for ambient intelligence, and one of its sub-projects, notably, is energy management in buildings.

The architecture proposed is made up of:

– sensors (temperature, humidity, light intensity, etc.) and actuators (power cuts to electrical sockets), controlled by low-consumption microcontrollers. A number of different low-consumption modes of communication (to avoid an excess energy cost) are possible depending on the installation: wireless, using IEEE 802.15.4; CPL (Watteco); or with an RS-485 industrial connection;

– interconnection gateway: typically located on each floor of a building, this gateway offers a Web interface whose configuration is intended for the infrastructure maintenance staff;

– back-office server: includes a management portal for the use of the building manager, a database for continuous measurements and a rules engine for adding automatic control loops (power cuts to certain inactive equipment, etc.).

OSAmI has a partnership with INEED, the *Centre d'innovation pour l'environnement et l'économie durables Rhône-Alpes* (Center for innovation for sustainable environment and economy) in France, which has its headquarters in a high "environmental-quality" building, and is a center of expertise on ecoconstruction.

In the context of this collaboration, OSAmI designed a sensor-based infrastructure which satisfies the needs of energy supervision in the building. The solution was then validated by way of pilot deployments in experimental houses, constructed in order to test new ecoconstruction techniques. In these in-situ tests, OSAmI's solution enabled the theoretical results about the properties of new materials to be validated experimentally for the first time. For instance, sensors located inside, outside and within walls made of hempcrete (hemp concrete) showed the good thermal inertia of this material, which is better able to keep ambient heat in once night falls.

The proliferation of technologies related to communicating objects – communications standards as well as open-source software platforms – will change the way in which we design and manage energy consumption in buildings. However, the occupants must become aware of the extent of their use of energy, in order to achieve an actual reduction in the energy footprint.

9.4. Conclusion

In this chapter, we have discussed the application of green networking in the context of the infrastructures of a smart city. We have illustrated a set of techniques which can be implemented to optimize the usage of embedded sensor systems, i.e. to reduce the number or the consumption of the devices.

Then, we touched upon the importance of norms in guaranteeing the durability of the technologies and presented an example of the use of these technologies to cater for the need for power management in buildings.

Green networking issues are at the heart of research in networking today and of the societal problems we find ourselves facing. A great deal of work still needs to be done in order to reduce the consumption of the protocols and sensor systems consistently, given that certain mechanisms can interfere with one another. Experiments, such as that conducted by OSAmI and discussed above, or by SmartSantander [SMA], are also interesting in order to properly comprehend the application of these results in a real-world environment.

9.5. Bibliography

[BAM 10] BAMBAGINI M., Power management in real-time embedded system, Masters Thesis, University of Pisa, 2010.

[BEN 00] BENINI L., BOGLIOLO A., DE MICHELI G., "A survey of design techniques for system-level dynamic power management", *IEEE Transaction on Very Large Scale Integration (VLSI) Systems*, vol. 8, no. 3, June 2000.

[BIL 11] BILAVARN S., RODRIGUEZ L., CASTAGNETTI A., "A video monitoring application for wireless sensor networks using IEEE 802.15.4", *Proc. 2nd Workshop on Ultra-Low Power Sensor Networks, WUPS 2011*, Como, Italy, 23 February 2011.

[GAY 11] GAY V., LEGUAY J., FERRARI G., MEDAGLIANI P., Procédé et dispositif de configuration d'un réseau de capteurs sans fils déposés, patent ref. 10006_BFF10P058, patent filed in April 2010 at INPI, extended in 2011 to the United States and Israel.

[GOL 95] GOLDING R., BOSH P., WILKES J., "Idleness is not sloth", *Proc. USENIX Winter Conf.*, New Orleans, 1995.

[ITE 11] ITEA2 Geodes Power saving handbook, http://geodes.ict.tuwien.ac.at/PowerSavingHandbook, 2011.

[KOP 11] KOPMEINERS R., KING P., FRY J., LILLEYMAN J., LANCASHIRE S., MING F., GROSSETETE P., VASSEUR J.P., GILLMORE M.K, DÉJEAN N., MOHLER D., STUEBING G., HAEMELINCK S., TOURANCHEAU B., POPA D., JETCHEVA J., SHAVER D., CHAUVENE C., "A standardized and flexible IPv6 architecture for field area networks, smart grid last mile infrastructure", http://www.cisco.com/web/strategy/docs/energy/ip_arch_sg_wp.pdf, December 2011.

[LU 00] LU Y.H., BENINI L., DE MICHELI G., "Operating system directed power reduction", *Proc. Int. Symp. Low Power Electronics Design*, Rapallo, Italy, July 2000.

[LU 04] LU G., KRISHNAMACHARI B., RAGHAVENDRA C.S., "An adaptive energy-efficient and low-latency mac for data gathering in wireless sensor networks", *Proc. Parallel and Distributed Processing Symposium*, Santa Fe, New Mexico, United States, 2004.

[MED] MEDAGLIANI P., FERRARI G., GAY V., LEGUAY J., "Cross-layer design and analysis of WSN-based mobile target detection systems", *Elsevier Ad Hoc Networks*, Special Issue on Cross-Layer Design in Ad Hoc and Sensor Networks, forthcoming.

[SUL 97] SULEIMAN D.R., IBRAHIM M.A., HAMARASH I.L., "Dynamic voltage frequency scaling for microprocessors power and energy reduction", *Electrical and Electronics Engineering (ELECO) conference*, Bursa, Turkey, 2005.

[ZAF 05] ZAFALON R., BACCHETTA P., "RT-OS run time power management for mobile terminals", *Embedded Systems Conference*, San Francisco, United States, March 2005.

Websites

[AVR] http://citavroraz.sourceforge.net.

[CAL] http:// www.ict-calipso.eu.

[M2M] http://www.etsi.org/website/technologies/m2m.aspx.

[MAC] The MAC Alphabet Soup served in Wireless Sensor Networks http://www.st.ewi.tudelft.nl/~koen/MACsoup.

[MIC] MicaZ, http://www.openautomation.net/uploadsproductos/micaz_datasheet.pdf.

[OPE] http://en.wikipedia.org/wiki/Open_data.

[OPE b] http://www.openmote.com.

[OSA] http://www.itea-osami.org.

[ROL] https://datatracker.ietf.org/wg/roll/charter.

[RUN] Runtime Power Management, Linux Weekly News, http://lwn.net/Articles/347573.

[SMA] http://www.smartsantander.eu.

[TEL] TelosB, http://www.willow.co.uk/TelosB_Datasheet.pdf.

[TIN] http://www.tinyos.net.

List of Authors

Aruna Prem BIANZINO
Politecnico di Torino
Turin
Italy

Lilian BOSSUET
Hubert Curien Laboratory
Jean Monnet University
Saint-Etienne
France

Florian BROEKAERT
Thales Communications & Security
Colombes
France

Tijani CHAHED
Institut Mines-Telecom
Telecom Sud Paris
Evry
France

Claude CHAUDET
Institut Mines-Telecom
Telecom Paris Tech
Paris
France

Salah Eddine ELAYOUBI
Orange Labs
Issy-les-Moulineaux
France

Vincent GAY
Thales Communications & Security
Colombes
France

Hicham KHALIFÉ
Thales Communications & Security
Colombes
France

Francine KRIEF
LaBRI
University of Bordeaux
France

Laurent LEFÈVRE
INRIA
Lyon
France

Jérémie LEGUAY
Thales Communications & Security
Colombes
France

Mario LOPEZ RAMOS
Thales Communications & Security
Colombes
France

Maïssa MBAYE
Gaston Berger University
Saint-Louis
Senegal

Paolo MEDAGLIANI
Università degli Studi di Parma
Parma
Italy

Martin PERES
LaBRI
University of Bordeaux
France

Jean-Marc PIERSON
IRIT
Paul Sabatier University
Toulouse
France

Guy PUJOLLE
LIP6
UPMC
Paris
France

Dario ROSSI
Institut Mines-Telecom
Telecom Paris Tech
Paris
France

Jean-Louis ROUGIER
LTCI
Telecom Paris Tech
Paris
France

Louai SAKER
Institut Mines-Telecom
Telecom Sud Paris
Evry
France

Sami TABBANE
Sup'Com
Tunis
Tunisia

Index